Child Prc
Se
Modern Mafia

Federally Financed Perjury, Fraud, Kidnapping, and Child Drugging for Profit

A Free Source Book
By
Dr. Eric D. Keefer, D.D

Dedication

This is dedicated to every child and family that has been victimized by CPS-DFCS agencies and every child who has died and been abused while in their unlawful custody.

It takes a village to raise a child, not a corrupt, immoral, and unethical "agency".

Mitacuye Oyasin

CHAPTERS

1. Introduction and Explanation
   Page 9
2. What to Do When CPS-DFCS Knocks on Your Door
   Page 12
3. How Exactly are CPS Caseworkers Qualified and Trained
   Page 40
4. The Process Used to Kid-Jack our Children
   Page 50
5. What About the Rules, the Law?   Page 61
6. Financial Incentives for the System   Page 65
7. Whom Can You Complain To?   Page 74
8. THE CASES   Page 81
9. The Corrupt Business of CPS   Page 96
   INDEX I   Page 114
   INDEX II   Page 116

Introduction

Kidnapping, coercion, fraud, perjury, and extortion occur in only the first 24-72 hours. This sounds like an action of a guerrilla group in Central America or perhaps brings back memories of the mobsters portrayed in the movies. However, this is in fact a government agency established for the welfare and protection of families and children.

The majority of cases that involve the removal of children from their home have shifted to become "removal without proper cause" or "lack of reason to believe any harm will come to the child". This occasionally done with good intent and if an attorney not appointed by the courts is involved, there is usually the result of an immediate return of the child. There are even more rare instances where circumstances are cleared up and the cases are released. If the family has resources and understanding to hire their own attorney, the likelihood of any type of case that is falsely started continuing beyond thirty days is miniscule.

The crimes have already begun with the unlawful removal of the child(ren) and continue unabated when the family is financially or educationally incapable of understanding and representing their own interests. Having had the child removed from their home without cause they mistakenly believe that their child will be returned to them because of this "mistake" that has been made. Even more abhorrent is the belief instilled in the parents that a "Family Court" appointed attorney will represent their best interests. This is almost never the case.

Fraud and perjury, intentional and criminal, is the next step in this process. The agency has to appear in court and prove their case. The way in which the courts function allow for fraudulent paperwork and perjury to occur unabated. The second fraudulent act occurs when they get the parents into their offices and tell them that "once the child has been taken we have no choice but to proceed". This is a lie that almost every parent will hear. This is done prior to or just after the court hearing and the parents, now desperate to see their child and traumatized by the kidnapping are told they must sign a "permanency plan" and work with the agency if they want to get their child back. The caseworkers are trained to be "compassionate and caring" during this con job in order to get the parents to sign.

By getting the grief-stricken, still in shock parents to sign this CONTRACT with the agency, it has managed to bypass the court as well as the laws and procedures set up to protect the parents from the fraud that is about to occur. The parents have now given up their autonomous rights to their child, and their right to contest any statements entered into the court as evidence by the agency. They also have forfeited their rights to contest the original, unlawful abduction of their child by the agency.

Coercion is the only true description of this process that can be applied. Using a human being to extort your demands from another, in this case traumatized relatives, is kidnapping for profit.

The fraud is increased when the parents have the misfortune of having attorneys appointed to them by the court system. In these cases, I have found not one example of the appointed attorneys informing the parents of their right to contest, about the contractual negation of that right, or defending them against the institution of an involuntary contract that they have been coerced into signing.

The system is set up in a way that makes it more profitable, measures success, and only rewards the individual parts if the child is made a ward of the state. The appointed attorneys get more pay the longer the process continues. The judicial part gets more funding by handling more cases. The child protection agency makes $7-15,000.oo per child per month. They all get nothing more when the cases are closed. They are not rewarded for reunification of the family, admission of errant action immediately, or preserving the rights of the child and parent. There is no money to be made by doing the right, the ethical, and the moral…and unfortunately these agencies know it.

The institution, CPS (DHS, CFS) established and supported by both the federal and state governments has been corrupted by the very processes they oversee. They have been allowed to act with impunity against the very purposes for which they were established. The agency, the courts, and the court appointed attorneys within this system are encouraged to do so by the very financial arrangements required to maintain their existence.

Fraud, coercion, extortion, and kidnapping are now the procedural norm? Intimidation along with vindictive or punitive action against parents who dare attempt to question the process or to stand up for their family's rights is "just part of the process"?

At least the Dons of old had to pay every individual judge, police officer, and official in order to get what they wanted, to break the law within limits. CPS-DFCS and the Family Court System in which they work are secreted away, self-investigating, and exist under laws and limitations of criminal prosecution that violate the laws and statutes of the states and the nation in which they exist. They pay themselves off and are able to act without limits.

This mafia, funded by the people of this nation, smiles and reassures you just like the Dons portrayed in the movies. The problem is, that like the old Dons, your fear of these agencies is real and their unchecked power is illegal. Under Title IV-B,

According to the Federal Rules and Regulations CPS-DFCS is responsible for the following:
1. To protect and promote the welfare of children (Section 425).
2. To prevent or remedy problems which may result in neglect, abuse, exploitation, or delinquency of children (Section 425).
3. **To prevent unnecessary separation of children from their families (Section 425).**
4. **To return children to their families after providing services to the children and the families (Section 425).**

5. To place children in suitable adoptive homes if it is not possible or appropriate to return the child to the family (Section 425).
6. To ensure adequate care of children in substitute care (Section 425).
7. To comply with reporting requirements for Title IV (Section 476).
8. To receive and spend child welfare earned funds for eligible child welfare services based on the Comprehensive Services Plan.
9. To qualify for additional IV-B funds by conducting an inventory of foster care children; operating a foster care information system, a foster care case review system, and a service program to reunite families or ensure other permanency for children; and implementing a program to prevent removal of children from their families (Section 427).

In this book, I am going to show you that not only do they completely disregard this list, but that they also have no financial incentive to abide by it. The money is made by acting against the very principles outlined to govern their actions and they do act against them with impunity.

Chapter One
An Accurate Explanation

I was approached in 2007 to do a book on the CPS-DFCS agencies nationwide and could not believe the information that the very well-known, outspoken, and reputable individuals who approached me were giving me. It was incredible, horrible, and truly unbelievable.

In fairness to myself, I must state that my wife had just passed away the December before and I was not in any state of mind to handle such an intricate task as correlation of evidence, interviewing for supplemental facts and data, nor to attack such a massive project in any way. However, with that understanding presented to you, the readers, I wish to formally apologize for my delay in addressing this issue.

In the time since I was approached, five years ago now, the corruption, fraud, deception, perjury, and any number of other UNCONSTITUTIONAL and illegal behaviors of this agency and the "Family-Juvenile" courts system across the United States has exponentially increased. The atrocities I am going to outline in this book are the stuff of horror movies, but can be easily found to be 100% accurate portrayals of the system. These stories are posted on television news websites, YouTube ©, and hundreds of web pages created by those who have been victims of the system.

I have heard many say that "those parents must have done something wrong or CPS/DFCS would not have been able to do anything like that". That is the rub. In the overwhelming majority of cases the families who have lost their children have done nothing wrong what-so-ever. The violation of their rights began when the Case Worker, untrained and unqualified to make subjective judgments about the state of parents or children, walks up the sidewalk in front of the house.

Those who are old enough will remember that "absolute power corrupts absolutely". In the case of CPS-DFCS agencies their power is absolute and is given to them in violation of both the laws of the states and the laws established by the federal government. It is not by chance that CPS, Child Protective Services agencies around the country are getting called the "Child Abduction Services" nor is it by chance that DFCS is now being labeled "Destruction of Families and Children Services".

I am focusing on the issues to be found in the Unites States only while writing this book. However, I want to emphatically state that many nations established such agencies on the same model with the same destructive powers and incentives during the 1980's. The U.S. allies such as Brittan, Australia, Canada, and others have the same processes and procedures in place with fewer or no constitutional protections for the parents and children who are victimized by the agencies in those nations.

As horrific, disgusting, and immoral as the actions in the U.S. agencies may be, all those who have been defrauded and violated by the U.S. CPS-DFCS agencies should be thankful that you are not in other nations which have no constitutional protections. You would be jailed for reading and spreading this text in violation of the orders of the closed and even more corrupt systems in place in those nations.

In the United States, these "Family Courts" and CPS agencies have attempted to ban or to confiscate manuscripts. In the United States they have attempted to create legal precedent in the "Family Court System" for the theft and banishment of intellectual property in order to keep their fraud and corruption hidden. In the case of attorneys or workers who have quit the system and become whistle-blowers they have been disbarred, blacklisted, harassed, threaten, and even unlawfully sent to jail for refusing to "cease discussing issues of the court in a public forum."

There have been verifiable, questionable deaths of individuals who have refused to remain quiet in several cases. Yet, they could not remain quiet while such atrocities are being funded and protected by the federal and state governments. There is nothing being remedied by either level of government because of the enormity of the profits and services payments that are being brought into the states and counties by acting illegally.

Chapter Two
What to Do if CPS-DFCS Knocks on Your Door…

This may seem to be out of order. However, there will be many people seeking just this kind of information because their child(ren) or grand-child(ren) have been taken by these agencies without cause. I am including all the information I am able and all the advice I have been able to accumulate.

The bare truth is that if CPS-DFCS knocks on your door you have a whole list of rights that **they have been trained to intimidate you into giving up**. They actually have classes that formally exist, to teach them how to "talk (their) way into a home". Then they have the "informal" classes based upon interaction of supervisors and caseworkers that are designed to teach workers how to INTIMIDATE their way into your home.

(A quick question, if they were acting in a legitimate way, why would they even need to convey such "intimidation" techniques to case workers as a matter of conducting business? Answer: THEY WOULDN'T and that just goes to show the extent of the corruption and disgusting nature of their intent and methodology.)

The first thing you need to know about CPS-DFCS workers showing up at your home is that they have absolutely no right to enter your home without a warrant. A warrant requires that they show probable cause to enter and inspect your home. If they use false premises to get the warrant, it is a crime that is prosecutable and creates the right to civil suit for violation of your rights. They get a warrant under false premises…they go to jail and pay you in civil court for their indiscretions.

The second thing is that you "keep your mouth shut". Don't volunteer anything to the CPS-DFCS worker who shows up at your door. If you have done nothing wrong, you have nothing to say that they want to hear.

NOTE that anything you do say will be twisted, warped, and misquoted to be used against you in a court room where you will not be allowed to hear the evidence against you. You also will not be allowed to face and question your accuser(s). You also will not be allowed to give evidence to the contrary, nor will you be able to contest any statements made by the case worker whether fraudulent or not. The reason you will not be able to do any of those things afforded as rights of due process is because neither you, your lawyer, nor any representative of your interests will be allowed in that courtroom.

The initial hearing will be in a "Family Court" (oxymoron) with only a judge, the case worker, the CPS-DFCS lawyer, and possibly a CPS-DFCS supervisor. No recordings will be made in over 90% courts, no transcripts will be available to you or your attorney. You will be given no opportunity to refute the claims before they come to your home and take your children.

This is done on the word of individuals who are working for an agency and within a system that can only gain financially by taking your child from you and keeping them under "state custody" for as a long a period as possible.

Based upon the testimony of several former CPS-DFCS case workers, no evidence of wrong doing, danger to the child, or other situations needing removal for remedy is required. Only the well-rehearsed testimony of a case worker, coached by the agency's attorney, is needed to abduct a child or keep a child from a home that has done nothing to warrant the removal of the child.

You should remain calm and polite during your interaction. Your only goal is to find out the accusations that have been made against you and to remain calm and polite. Did I mention that you need to remain calm and polite? This person has no rights at this point under the law, but based upon the overall behavior of CPS-DFCS case workers, it is in your best interest to handle them in the most polite and appropriate manner possible.

The vindictiveness, false testimony, perjury, and fraudulent statements made by case workers is well-documented and you must assume that if the person in front of you is working for these agencies, their moral and ethical character is questionable at best and non-existent if they have been with the agency more than a year.

The third step is to request that the case worker present their warrant to you. Under the 4th Amendment of the Constitution in the United States you are protected from violation of your privacy in your home. The following material from a caseworker training program displays that they know they are violating your rights when they attempt to harass, intimidate, or threaten their way into your home.

I have included the "cases and court rulings" as well, because this proves that they are trained and any "counter-training" designed to defraud you of your rights is knowingly done against the mandates of Senate, as well as violates the agency rules and the requirements of the law.

> Federal law now requires all child protective services workers ("CPS workers") to be trained about the legal rights of parents. The United States Senate explained why it added this section to the Keeping Children and Families Safe Act of 2003:
>
> While the committee is strongly committed to the main mission of the child protective services system--to ensure that child safety and the best interests of the child are protected, the committee believes it is important for child protective services personnel to understand and respect fourth amendment limitations on their right to enter a home when investigating an allegation without a court order.
>
> One basic freedom is the right to privacy in the home. This particular right is enshrined in the Fourth Amendment to the U.S. Constitution, which is part of the Bill of Rights. This part of the Constitution says: " The right of the people to be secure in their persons, houses, papers, and effects, against unreasonable searches and seizures, shall not be violated... "

Congress wanted to make sure that every CPS worker is "fully aware of the extent and limits of their legal authority and the legal rights of parents" each time they carry out a child abuse or neglect investigation.

Reasonable Expectations of Privacy

Every American citizen has a right to a reasonable expectation of privacy in his or her home and family life. What counts as a "reasonable" expectation of privacy during a child abuse or neglect investigation? Some social workers have assumed that parents have no meaningful right to privacy under such circumstances, but the courts have disagreed.

Courts have ruled that the Fourth Amendment right to a reasonable expectation of privacy means that government workers must have one of three things before they can constitutionally search for evidence of wrongdoing in a private home.

They must either have (1) consent, (2) exigent circumstances, or (3) a warrant or other valid court order authorizing entry.

Consent — A parent who has a right to privacy in the home can waive that right and invite CPS workers in to conduct an investigation. The vast majority of CPS investigations take place with the consent of the parents. In order to be sure that an investigation is legal, CPS workers should ask the parents for permission to enter their home. Valid consent cannot be obtained through threats or lies.

Exigent Circumstances — If the parents won't let the government in, but there's a real emergency, special rules apply. Nobody has a reasonable expectation of privacy in an emergency situation. You don't need a permission to rescue a child from a burning building! This is considered an "exigent circumstance" and is an exception to the normal requirement that government workers need consent or a court order to enter a private home. If the allegations aren't serious, aren't credible, or the children don't appear to be in immediate physical danger, then the circumstances probably aren't "exigent."

Exigent circumstances are not just another way to carry out an investigation: for any circumstances that are truly "exigent" are extreme enough to halt the investigation until the safety of the child has been ensured. It is appropriate to take a child into temporary protective custody if that is truly necessary to ensure the child's safety, but temporary "protective" custody should not be used as temporary "investigative" custody. CPS workers take a grave legal risk whenever they take a child into custody solely for the purpose of questioning the child, since that constitutes a "seizure" for purposes of Fourth Amendment law, and any such seizure must be justified at the outset.

<u>Court Order</u> — If the parents won't consent but there is no emergency, CPS workers need a court order to enter the home or seize a child. The Fourth Amendment authorizes search warrants, as long as there is "probable cause" to believe that a crime or other violation of the law has occurred or is occurring.

Why Respecting Rights Matters

There are many reasons CPS workers need to understand the rights of parents. First, of course, the Constitution requires it. That should be enough for any American citizen. But there are three good practical reasons why respecting parental rights improves the child protection system.

Winning Cooperation through Respect — It's easy for CPS workers to wind up in court if parents won't let them in their home and refuse to allow them to interview the children. If a government employee forces his or her way into a home without consent or exigent circumstances, they can be sued for civil rights violations.

If they don't force their way into the home, they need a court order to get in. Everyone is better off if CPS workers go to court to get the authority they need, rather than acting without constitutional authority and getting sued for it. People fear what they don't understand. Frightened parents often make things worse for their families out of ignorance. It takes some time to help parents understand how the system works and to help them understand their rights, but this small investment of time up front can save a great deal of time in litigation later. By demonstrating a real respect for their rights, CPS workers can improve the chances that they will cooperate.

Building Effective Relationships — Even if the parents insist on going to court, the chances of helping the children are better if CPS workers demonstrate their respect for the parents throughout the process. If CPS treats the parents with dignity before, during, and after court proceedings, the family is much more likely to work constructively with CPS when and if the court authorizes intervention.

Avoiding Civil Rights Liability — Last, but not least, the chance of losing a civil rights lawsuit drops to zero if CPS workers show that they understand

and respect the rights of parents at every step of the investigation process.

Case Study: Calabretta v. Floyd
Facts: On Thursday morning, an anonymous caller alleged that she had once been "awakened by a child screaming 'No Daddy, no' at 1:30 a.m. that morning at the Calabretta home." She said she had also heard one of the children scream 'No, no, no' two days before. Social worker Jill Floyd did not go out to investigate until four days later. Mrs. Calabretta refused to let Jill in, but the children were standing at their mother's side and did not appear abused or neglected.

Ten days later, Jill returned to the Calabretta house with a policeman. Mrs. Calabretta did not open the door, and said she felt uncomfortable letting them in without her husband at home. The policeman said if she did not open the door, he would force his way in. Mrs. Calabretta then opened the door and let them in.

Was the entry into the home permissible?
    Analysis: A constitutional entry into a private home requires consent, exigent circumstances, or court order. There was no court order in this case, so the only options are consent or exigent circumstances. The federal court that heard this case concluded that a reasonable jury could decide that Mrs. Calabretta did not consent to this entry, even though she opened the door. The police and social workers argued that they were entitled to enter the home under "exigent circumstances" because they were investigating the possibility of harm to a child.

    Ruling: The Ninth Circuit Court of Appeals ruled against the social worker and police officer. "The facts in this case are noteworthy for the absence of emergency. The social worker and her department delayed entry into the home for fourteen days after the report, because they perceived no immediate danger of serious harm to the children." Calabretta v. Floyd, 189 F.3d 808, 813 (9th Cir. 1999) [emphasis supplied].

Case Study: Good v. Dauphin

Facts: An anonymous caller claimed that seven-year-old Jochebed Good had bruises on her body and that Jochebed said some of the bruises were caused by a fight with her mother. She didn't go to school the next day, so at 10 p.m. that night, Mrs. Good was startled by loud pounding at her door. There she saw a uniformed police officer and a man from Social Services who said, "You must let us see your daughter."

Mrs. Good asked to see a warrant or a court order, but the social worker said they did not need a warrant and that they had a report that her daughter had been abused and she must let them enter. Mrs. Good wanted to telephone a lawyer, but it was too late at night. The police officer called the station on her walkie-talkie and informed the duty officer that "they were going in" to the home. At this point, Mrs. Good allowed them in, but said she did so only because she understood that she was being compelled to do so.

Was the entry into the home permissible?

Analysis: A constitutional entry into a private home requires consent, exigent circumstances, or court order. The police and social worker said they did not need a warrant, so the only options are consent or exigent circumstances. The federal court that heard this case concluded that a reasonable jury could decide that Mrs. Good did not consent to this entry, even though she opened the door. The police and social workers argued that they were entitled to enter the home under "exigent circumstances" because they were investigating the possibility of harm to a child.

Ruling: The Third Circuit Court of Appeals refused to treat this case as an emergency situation. The anonymous report merely said that Jochebed once "had bruises on her body" of unspecified severity. Nothing the officials saw or heard nothing during their conversation with Ms. Good at the door suggested in any way that Jochebed was being mistreated. The court said, "The [exigent circumstances] exception must not be permitted to swallow the rule: in the absence of a showing of true necessity—that is, an imminent and substantial threat to life, health, or property—the constitutionally guaranteed right to privacy must prevail." Good v. Dauphin County Social Services for Children and Youth, 891 F.2d 1087, 1094 (3rd Cir. 1989) [emphasis supplied].

Credible Evidence
Federal law requires CPS to investigate every allegation that meets certain specific criteria, but not all allegations are credible. In fact, as any experienced CPS worker knows all too well, some allegations are malicious. The most malicious reports tend to allege horrific abuse which would clearly constitute "exigent circumstances" if true. CPS workers should be careful to look for corroborating evidence before taking steps that can frighten children and violate parents' constitutional rights, possibly leading to civil rights litigation.

All three cases examined above involved anonymous reports. Courts treat anonymous tips as significantly less credible than reports from identifiable individuals. Judges will not issue a search warrant for an anonymous tip unless there is significant "corroborating" evidence to support it. When authorities force a warrantless entry solely on the basis of an uncorroborated anonymous tip, however, they take a great risk. While the tip might be accurate, it might just as well be false. Overzealous officials can terrify children, traumatize parents, and wind up being sued for violating civil rights. The risk of error and the resulting harm to family and children are too great to ever justify entering a home on the basis of an uncorroborated anonymous tip.

In the Boggess case, the Wisconsin Supreme Court looked hard at the credibility of the anonymous call. Before the authorities entered the Boggess home, they had no way to be certain that the report was accurate. Here's what they found when they entered, however:

Once inside the home, Hammel went over to L.S. and saw that a pronounced part of his lip was missing and that the wound was inflamed and needed to be cleaned. Without directing the question to anyone in particular, Hammel asked, "What happened? How did he get hurt?" Calvin Boggess responded that he had fallen on L.S. and had hurt him. Later, Boggess stated, without prompting, that he had spanked both children several times.

With Janice Boggess present, Hammel examined L.S. more thoroughly in a rear bedroom. She observed that he had bruises on both sides of his legs from the ankles to the thighs, and that his arms were black and blue from the elbows to the wrists and halfway up his back. She also noticed that L.S. had hair missing from the top of his head, and that he walked with a "waddled limp." Hammel then examined K.S. and observed bruises on her body.

Calvin Boggess was tried for felony child abuse. He tried to exclude all this evidence from his criminal case because it was obtained without a warrant. If the court had excluded the evidence—and it was a close question—this criminal might have gone free.

CPS workers face twin risks every time they enter a home without a warrant. If the parents are innocent, children get frightened, parents get angry, and social workers get sued. If the parents are guilty, however, social workers take the risk of letting a criminal child abuser off on a "technicality." The right answer is to get a warrant or court order before coercing entry into a home on anything less urgent than an emergency.

Case Study: Wallis v. Spencer

Facts: Bill and Becky Wallis were happily married with two children, but Becky's sister Rachel had mental problems. The Wallises cut off all contact with Rachel after she falsely accused Bill of sexually abusing his little girl. The next year, Rachel was hospitalized for suicidal paranoia and multiple personality disorder. In the hospital, she told her therapist about a recently recovered "memory."

She said that she and her father were in the woods twenty years earlier, where he wore a cultic robe and hypnotically chanted, "On the third full moon after two blue moons a child will be killed." One of Rachel's "alter egos" then told the therapist that the "child" in the twenty-year-old incantation was her two-year-old nephew Jessie, and that her brother-in-law Bill intended to sacrifice Jessie to Satan at the "Fall Equinox" ritual in a few days.

Based on this information, social workers told police officers to "pick up" the children. The police seized both children without consent or court order.

Was the seizure of the children permissible?
Analysis: A constitutional seizure of a child requires consent, exigent circumstances, or court order. The police argued that there were "exigent circumstances" because they were trying to prevent a human sacrifice.

Ruling: The Ninth Circuit ruled in favor of the parents. "Officials may remove a child from the custody of its parent without prior judicial authorization only if the information they possess at the time of the seizure is such as provides reasonable cause to believe that the child is in imminent danger of serious bodily injury and that the scope of the intrusion is reasonably necessary to avert that specific injury." Wallis v. Spencer, 202 F.3d 1126, 1138 (9th Cir. 2000) [emphasis supplied]. The court ruled that a mental patient's delusions were not a credible basis for a "reasonable" seizure.

Case Study: H.R. v. State Dept. of Human Resources

Facts: Alabama CPS workers received an anonymous report of child abuse or neglect regarding Helene Richards's four children. A second report came in the next day. The evidence she offered the court consisted of the following anonymous allegations: two children being kept shut in a room; a child being sick and not receiving sufficient food; a child choking a cat; a child trying to "hang" himself; the children being kept in the back of a van; and infestations of roaches in the physical home of the children.

CPS worker Donna Jones was assigned the case four weeks after the anonymous reports came in. She did not go out to the home for another seven days. When she got there, the mother kept her outside and would not let her interview the children privately. Ms. Jones filed a petition to gain access, based on a statute that authorizes Alabama courts to issue such an order "upon cause shown."

Analysis: A constitutional entry requires consent, exigent circumstances, or a court order. Judges will not issue a search warrant without "probable cause," which means credible evidence that a crime or other violation of the law has been committed or is being committed. An uncorroborated anonymous tip falls short of the "probable cause" standard. CPS argued that Alabama law only required "cause shown," not "probable cause."

Ruling: The Alabama Court of Appeals ruled in favor of the mother. The court wrote, '[T]he power of the courts to permit invasions of the privacy protected by our federal and state constitutions, is not to be exercised except upon a showing of reasonable or probable cause to believe that a crime is being or is about to be committed or a valid regulation is being or is about to be violated." H.R. v. State Dept. of Human Resources, 612 So.2d 477, 479 (Ala. Civ. App., 1992) [emphasis supplied]. The court refused to rely on uncorroborated hearsay to invade the privacy of a home.

Interviewing Children--Most states expressly authorize CPS workers to interview children without a parent's permission. This means CPS workers don't need to get the parent's consent to talk to a child at a public school or other location where the child is temporarily in the custody of some other adult who consents to the interview. It does not empower government workers to forcibly separate the child from his or her parent for the purpose of an interview.

That is a "seizure," for constitutional purposes, which requires consent, exigent circumstances, or a court order.
(I would emphatically state that any interview with a child outside of the parent's custody and without informing them is a violation of their rights as well as the child's. An "independent" attorney or the parent's attorney should be required to protect both the child and the parents from manipulation. Any caseworker who does this after being denied in the home is reaching the limit of acceptable action at the least and approaching gross misconduct at the worst. That they do this with no witnesses and no limitations should be of great concern considering how many false accusations occur after such interviews.)

Court after court has determined that the test of whether a government action is a "seizure" or not is whether the subject of that action feels free to leave. If a child feels like he can leave the interview at any time, he has not been "seized." If, on the other hand, the child does not feel free to go, the Fourth Amendment applies.

Here is how to analyze an interview with a child:
If a mature child indicates a willingness to talk to a government worker privately, the child's consent may satisfy the Fourth Amendment. Younger children, however, are too impressionable to give meaningful consent. If a child is not old enough to babysit, the child is probably not old enough to waive his or her constitutional rights.
If the parent or other caretaker allows a government worker to interview the child, there

may be a "seizure," but it is constitutional because consent was given.

If there is evidence that the child is in immediate danger, a government worker may seize the child because of exigent circumstances.

In all other cases, government workers need authorization from a court before they can separate a child from his or her parents for the purpose of an investigation.

Many states have statutes that allow CPS workers to interview a child without the parent's permission. These statutes have been upheld as constitutional for interviews at public schools, but each public school student is in the care of a public school employee who is required by law to consent to these investigations. (This allows a school employee to consent in the place of a parent??? Fishy, sneaky, and clever, but legal???) Cases involving children at private schools have had very different outcomes.

Medical Tests: The statutes that allow CPS workers to interview a child without parental consent do not automatically empower CPS workers to order x-rays or other medical tests. In the absence of a statute to the contrary, these are medical decisions that fall within the scope of traditional parental rights. [5] A child or a babysitter can choose whether or not to let the child talk to a stranger, but neither a child nor a babysitter can consent to invasive medical procedures. Only a parent, a legislature, or a court can authorize such actions in the absence of a medical emergency.

Case Study: Doe v. Heck

Facts: CPS workers received a letter alleging that a ten year old girl had been bruised by a spanking she received at the Greendale Baptist Church and Academy. CPS worker Carla Heck went to the school to investigate. Despite the objections of the principal, and without parental notice or consent, Ms. Heck removed eleven-year-old John Doe Jr. from his fourth-grade classroom and interviewed him about corporal punishment. Heck relied on Wisconsin statute § 48.981, which said: " The agency may contact, observe or interview the child at any location without permission from the child's parent, guardian or legal custodian if necessary to determine if the child is in need of protection or services, except that the person making the investigation may enter a child's dwelling only with permission from the child's parent, guardian or legal custodian or after obtaining a court order to do so". The child, parents, and private school sued Heck and others for civil rights violations.

Analysis: The CPS worker relied on a Wisconsin statute that expressly authorized her to talk to children without permission from the parents. The United States Constitution, however, prohibits unreasonable searches and seizures. May a CPS worker rely on a statute that authorizes search and/or seizure on private property without a warrant, consent, or exigent circumstances?

Ruling: The Seventh Circuit Court of Appeals struck down the statute but did not hold the CPS worker personally liable for relying on it. "[T]o the extent § 48.981(3)(c)1 authorizes government officials to conduct an investigation of child abuse on private property without a warrant or probable cause, consent, or exigent circumstances, the statute is unconstitutional." Doe v. Heck, 327 F.3d 492, 515 - 516 (7th Cir. 2003) [emphasis supplied].

Frequently Asked Questions

Q: My investigation protocol requires me to visit the home and interview each child privately. The parents refuse to let me in and won't let me talk with the children. What should I do?

A: Make sure you have the latest version of the protocol. Many states used to insist on private interviews with all children in all cases without any concern for the Constitution. If your investigation protocol still requires this, please report it to an attorney for your agency.

Q: I thought police and CPS workers operated under different rules. What has changed?

A: Police and CPS workers are both government workers, and are therefore both ultimately governed by the Constitution. The "different rules" that apply to police and CPS workers are the result of different court systems. Police routinely testify in criminal courts, which exclude all evidence that has been unconstitutionally obtained. As a result, police are very careful to make sure they comply with the Constitution at all times. CPS workers usually testify in juvenile courts, which allow the use of evidence that has been unconstitutionally obtained. Because the juvenile courts allow this evidence, CPS workers may not know they have violated the Constitution until they are personally sued for civil rights violations. Congress now requires all CPS workers to be trained about parents' rights.

Q: Isn't there a "special needs" exception to the Fourth Amendment that allows CPS workers to protect children?

A: The United States Supreme Court has identified certain "special needs" situations where traditional Fourth Amendment restrictions do not apply, but the "specials needs" exemption does not apply when the government is investigating allegations of unlawful activity. As the Tenth Circuit recently said, "The Fourth Amendment protects the right of the people to be 'secure in their persons' from government intrusion, whether the threat to privacy arises from a policeman or a Head Start administrator. There is no 'social worker' exception to the Fourth Amendment." Dubbs v. Head Start, Inc., 336 F.3d 1194, 1205 (10th Cir. 2003).

Q: Why can Welfare workers enter a home without a warrant but CPS workers cannot?
A: The United States Supreme Court has ruled that Welfare workers do not need a warrant to enter a private home. Wyman v. James, 400 U.S. 309 (1971). Subsequent cases have made it clear that this is because the penalty for refusing to allow a Welfare worker into the home is a denial of future Welfare benefits, not removal of a child or criminal prosecution.

Q: How can I be sure the circumstances are really "exigent"?
A: This is an uncertain area of law, which means CPS workers should generally try to get a court order if there is time to do so. The Second Circuit Court of Appeals has decided that CPS workers in Connecticut, New York, and Vermont must get a court order if there is time to do so. Tenenbaum v. Williams, 193 F.3d 581 (2nd Cir. 1999).

The Tenth Circuit, which covers Colorado, Kansas, Oklahoma, Nevada, New Mexico, and Utah, agrees. "Simply put, unless the child is in imminent danger, there is no reason that it is impracticable to obtain a warrant before social workers remove a child from the home." Roska ex rel. Roska v. Peterson, 328 F.3d 1230, 1242 (10th Cir. 2003).

The rule is different in Alabama, Florida, and Georgia, because the Eleventh Circuit Court of

Appeal has ruled that CPS workers do not need a court order to remove a child if the child really is in immediate danger. Doe v. Kearney, 329 F.3d 1286 (11th Cir. 2003).

That rule can be risky, however, as the Ninth Circuit shows. A CPS worker who removed a child from her home without a warrant could be sued for doing so without truly exigent circumstances. Mabe v. San Bernardino County, Dept. of Public Social Services, 237 F.3d 1101 (9th Cir. 2001).
Given this disagreement between the circuits, it is just a matter of time before this question winds up before the United States Supreme Court.
(This is actually funny. Keep in mind this is written for caseworkers, and note that there is no disagreement among the courts what-so-ever. They are all emphatically stating that the CPS workers should come to the court and get their paperwork right before abducting a child and to do so is to do so at their own risk. They all say the same thing; TRULY IMMEDIATE DANGER is a requirement that must be met, period.)

Q: Doesn't it make a difference that the CPS worker is trying to protect children, not punish parents? (that is a loaded and inaccurate statement, they do not protect children and are vindictive toward any parent who will not cooperate with unsubstantiated claims made by caseworkers. In many instances the county agency has attacked other family members and manufactured cases against them in retaliation.)
   A: The Fourth Amendment guarantees the privacy of the home, regardless of the government's motives. The Tenth Circuit Court of Appeals put it well, saying, "[T]he defendant's motive to protect the child ... does not vitiate plaintiffs' [constitutional] rights. That motive, however, may

enter the calculus of the damages, if any, that his actions justify." Roska ex rel. Roska v. Peterson, 328 F.3d 1230, 1255 (10th Cir. 2003).

Conclusion: The Constitution Is Good For Children

CPS workers who are unfamiliar with constitutional law may wonder whether all this emphasis on parental rights is good for children. There are a number of reasons why children are better off when CPS workers demonstrate their respect for family freedoms.

Respect puts criminal child abusers in jail. It's one thing to pull a child out of a dangerous situation, but far too many criminal child abusers go on to hurt other children. Remember the Boggess case! An abuser may lose his own children and still move into some other home where children are present. Constitutionally admissible evidence puts abusers in jail. That is good for children.

Respect builds trust. Treating parents with dignity makes it easier for them to accept help if it is needed. If CPS workers tell parents, "You must let me in or I will take your children away," they may submit but they will never trust CPS again. If CPS workers say, "You have a real choice in this matter," the parents are much more likely to trust CPS in the future, which means they are more likely to get the help they need to succeed as parents. That is good for children.

Respect protects innocent children and families. Too many people have learned how to use anonymous allegations of child abuse as a way to attack innocent families. Some of the most awful allegations of abuse are also the most false—and these false allegations hurt innocent children and families. Weeding out false and malicious allegations is good for children.

> Respect protects CPS workers. CPS workers who do not know the law tend to break it. There is no excuse for reckless disregard of federally protected civil rights: a zeal for protecting children without proper respect for family freedoms can subject CPS to civil rights suits and substantial damages.
> The best place for a child is in his or her own home, in the care of fit and loving parents. Most parents believe they are both fit and loving, even if their behavior puts their children at risk. CPS workers can do a better job of protecting children if they follow the Constitution in each investigation of abuse or neglect. Whether the investigation winds up clearing the family of unfounded allegations or helping parents get the services they need, the CPS worker's attitude of respect makes a real difference. The Constitution doesn't keep CPS workers from doing their job. It makes that job possible. Our Constitution is designed to enable the government to punish the guilty and protect the innocent. Nobody deserves punishment more than criminal child abusers, and nobody deserves protection more than innocent children.

So ends their training paper, but what is the truth in comparison with this document and training class created to meet the demands of the U.S. Senate? Well, they teach the training classes as the Senate instructed. Every caseworker is required to know this.

Then they have seminars and "informal" instructional days in which they teach the caseworkers how to circumvent this whole thing, to "create" exigent circumstances, and to enter the home/remove the child against the agency's protocol, against the direction of the federal district courts, and against the clear delineations made by the Senate.

It is common for toys left on the floor, dust on shelves, garbage not taken out yet, and other minor issues of living in a home with children to be misrepresented as homes being unkempt or inappropriately maintained. I should point out it doesn't matter that they do any of this (illegally, fraudulently) because the court system that CPS-DFCS works within is set up to protect them and the procedures are by their very nature enough to work against any parents wrongfully accused or perjured against in "Family Court" by a caseworker.

Now that you understand that they know they are intentionally attempting to intimidate and coerce their way into your home, intent to violate your 4th Amendment rights after having been informed that they ARE NOT TO DO SO by ruling after ruling by circuit judges, we move on to step four.

DO NOT SIGN ANYTHING without speaking to an attorney. Again, in the "informal" classes at the CPS-DFCS agencies, caseworkers are trained to intimidate, misrepresent their authority, and to intentionally misrepresent your legal rights in order to obtain your cooperation in the removal of your child.

They are going to try to get you to sign a release for them to contact physicians, teachers, the school, and any other agent that may be involved in the care of your children. They will say that you "have to", "must", "are required to", and any other number of lies because they are taught to do so, but again I emphatically warn…DO NOT SIGN ANYTHING WITHOUT YOUR ATTORNEY.

Step five is to immediately begin documenting everything. It doesn't matter if you did nothing wrong. As a matter of fact most parents whose children are in CPS cases are parents who also did nothing wrong. You have no reason to believe, trust, or expect any honesty, integrity, morality, or ethics from the CPS-DFCS workers, supervisors, or the family court attorneys and judges.

The last ten years has proven that those traits are not part of the system. If those positive character traits existed within the system they would interfere with the federal payments received by all the parties, disrupt the service industries that make billions each year from defrauding parents, and prevent all parties involved from getting their bonuses that come from unlawful abduction of children, fraudulent court documentation, perjury on the part of CPS-DFCS workers.

You must document everything from the moment CPS-DFCS begins their process. You ABSOLUTELY must write down the dates, times, names, and agencies for every person you speak with in person or by phone. Along with each entry you must also be sure that all information in the conversation is documented as well. Keep all the paperwork you receive and organize it by date.

Immediately request any transcript from the original "closed-door" hearing if it happens and do not let your lawyer off with regard to this important document. When they fail to present it, then you have the grounds for your Federal complaint against CPS-DFCS, their attorney, and the judge who allowed the "closed-door" hearing on the grounds of unlawful trial and conviction.

They have testified behind closed doors and you now have less than 30 days to file a counter-claim and complaint of fraudulent documentation and perjury against the CPS worker, their attorney, and the family court judge. Take it to federal court because they have violated every law with regard to having a hearing against you.

Step Six and Seven are hand in hand. NEVER TRUST A SOCIAL SERVICE WORKER from CPS OR DFCS and you need to talk to an attorney as soon as possible. The last decade has shown that CPS-DFCS workers will lie, manufacture evidence, twist your words, and even forge documents because they are able to within the system that they operate. I again point out that no real evidence is required in the "initial hearing" and no representative of your interests is going to be allowed in the court.

They can and do say what they want, what is needed to draw your child into the system and after 30 days their testimony cannot be challenged. It is a violation of all kinds of laws and rights, but somehow the "Family Court" system has managed to establish their own rules and statutes of limitations designed specifically to allow this to occur and to limit a parent's ability to get their children back. If you do not challenge them within the first 30 days, you will never be able to do so in the family court system, you should immediately be going into the Federal Court System to file complaints against the agency, attorneys, and judge.

The keeping and recording of accurate and complete documentation is essential to allow you the opportunity to do so.

According to the Annie E. Casey Foundation, the removal from family is so devastating that within 4 hours, children experience 6 times more post traumatic stress than combat troops. It is widely acknowledged that foster children suffer four times more abuse and neglect that children in the general population.

The National Center on Child Abuse and Neglect reported in 1998 that six times as many children died in foster care than in the general public. The number is now that you are nearly 90% more likely to be abused in custody of CPS-DFCS or in an assigned Foster Care Home than if you were in the general public.

Additionally, foster children suffer severe life altering trauma from forced separation from family, depression, long term court ordered adult psychotropic medication for children as young as 3 years of age to cope with trauma after forced separation. The majority of aging out foster children have no family ties, no support system, are homeless within three months, suffer severe mental illness, illiteracy, unemployment, poverty, and incarceration.

Restoring the rule of law is especially critical in light of national systemic corruption in the provision of children welfare services. Social workers, more often than not, are guilty of misconduct, fraud and corruption. In Georgia, well respected Nancy Schaefer, former Republican Senator, 50th District, wrote a 2008 article,

"Kidjacked. The Corrupt Business of Child Protective Services". It is absolutely necessary to re

"I have witnessed such injustice and harm brought to families that I am not sure if I even believe reform of the system is possible! The system cannot be trusted. It does not serve the people. It obliterates families and children simply because it has the power to do so....It is time to pull back the curtain and set our children and families free."

In "Kidjacked ", Senator Schaefer stated: poor parents are targeted because they do not have the where-with-all to hire lawyers and fight the system; caseworkers and social workers are oftentimes guilty of fraud, withhold evidence, fabricate evidence; the separation of families is a growing business because local governments have grown accustomed to having taxpayer dollars to balance their ever expanding budgets; Child Protective Services and Juvenile Court hide behind confidentiality to protect their decisions and keep the funds flowing; parents are victimized by "the system" that makes a profit for holding children longer and "bonuses" for not returning children; state employees, lawyers, court investigators, court personnel, judges, psychologists, psychiatrists, counselors, caseworkers, therapists, foster parents, adoptive parents all look to the children in state custody to provide job security; social workers are the glue that holds "the system" together that funds the court, the child's attorney, and the multiple other jobs; the Adoption and Safe Families Act offers cash "bonuses" to the state for every child they adopt out of foster care; in order to receive the "adoption incentive bonuses", local child protective services need more children-more merchandise to sell; no one in the system steps up to the plate and says, "This must end!" because they are all in the system together.

Chapter Three
How are CPS Workers Deemed Qualified and Trained?

The short answer is that there are no real "set standards for the hiring of CPS (DFCS) caseworkers" and that the training that is formal within the agencies is much less important than that which is implied and "informally passed on from senior caseworkers and supervisors to the new arrivals".

CPS social workers are responsible for accessing, identifying and documenting cases of abuse or neglect among children and finding the appropriate services to enhance child welfare. CPS social workers must be able to identify immediate threats made to a child and provide arrangements that comply with state and federal laws and agency procedures.

These duties may include, but are not limited to providing counseling and support services to children and parents, referring children and families to other services if necessary, placing children in foster care, finding adoptive homes for children with no adult caretakers, and the discernment of conditions that may result in the removal of child from its family.

How about checking out the qualifications most CPS social workers have? The problem is that it is impossible to do. The requirement of a degree is typically in place or out of the workplace depending on the staffing needs of the agency. However, the degrees that they can possess if they have any at all vary greatly.

Some have degrees in social work (see Index A, Index B) while others have degrees in Early Education, Child Development, and even Spanish or irrelevant studies, and a third group has acquired their social work degree through an "online" college that requires only a "one-time" payment and a couple weeks in order to obtain an A.A..

The training programs for new hires vary from state to state and county agency to county agency. In Arizona, numerous CPS workers testified that they spent 3-4 days reading policy manuals and were assigned to follow "seasoned caseworkers for a short period of time. In New York City, caseworkers receive 20 days of training, most of which focuses not on child development but on filling out forms and other paperwork tasks. A panel of experts in the field of child welfare assembled to study the operations of the Illinois Department of Children and Family Services in the wake of a lawsuit against the agency. The panel found that: "DCFS caseworkers are uninformed or misinformed about internal resources and procedures of DCFS, as well as about DCFS philosophy and job performance expectations."

"Contrary to public opinion, the majority of CPS workers are not trained social workers," write Professors of social work Lela Costin, Howard Karger and David Stoesz. They note that reductions in the minimum educational standards for public service jobs, including child protective services positions, have become a national trend. Costin and colleagues point to a national study in which staff development directors in 27 states were asked about minimum educational requirements for child welfare caseworkers. Respondents reported that none of those states required a social work degree for an entry-level position in child welfare. Nor would supervisory staff necessarily be better prepared. In a study of front-line supervisors, it was found that while some had many years of experience, *barely one-third held graduate degrees in social work!*

Dr. Deborah G. Alicen, a clinical psychologist who wrote her doctoral dissertation on Child Protective Service reports that her research indicated that many caseworkers have no college training, and that only 28 percent of all CPS workers in the country have either a Bachelor of Science or a Master of Science in Social Work degree. Thus, 72 percent of them, she concludes, have no suitable professional training. A study of 5,000 child welfare workers would bear this out. Researchers found that only 15 percent held a Bachelor of Science degree in Social Work, while only 13 percent held a Master of Science degree in the field.

The lack of training often has a tremendous impact on how investigations are conducted, as well as their outcomes. Law enforcement personnel in the state of Florida note that there were many caseworkers who were accusatory in tone from the outset of the initial interview, attributing this to their lack of training, among some other factors.

A recent study indicates that some of the state's child protective services caseworkers were "notorious for their cold, almost confrontational, style thus putting families on the defensive. In the absence of proper training, some were unnecessarily and unwisely heavy handed and reacted to resistance with an exertion of power," note reviewers. As a result, the agency had a widely held reputation of being "an invasive bureaucracy which removes children from their families without good cause."

The lack of training and experience is not limited to the child protective services intake workers, extending throughout the entire child welfare bureaucracy. According to David Liederman, Executive Director of the Child Welfare League of America, "there is inadequate training for foster parents, for caseworkers, for core personnel, and for administrators. I would suggest to you that it is not in the best interest of children or families in this country to hire someone with a B.A. in history or art, give them three weeks of training, and turn them loose."

Is this what was intended by the original legislation? Not according to Pat Schroeder, former Congresswoman and sponsor of the Mondale Act, who writes, "In the Child Abuse and Protection Act, we tried to set up a system in which trained people could identify children and parents who were in trouble and offer them help. . . People's lives can hinge on the judgment of the social worker assigned to their case. Yet some states don't even require social workers to have specialized training in their field."

What about those who actually have "social work" degrees? What about them? They get a B.A. or A.A. degree that teaches them how the system operates and how to function within the system? What does that have to do with making decisions that can save or destroy the lives of children and whole families? Not a thing is the only answer anyone who can think for themselves should even think of.

What does knowing how the system works, how the procedures of the agencies are established and performed, and how to work within the system have to do with managing and making decisions about the lives and well-being of children and families? NOTHING. So, how are they qualified to do this kind of work with this degree? THEY ARE NOT and that is the only truthful answer that can be given.

Would you want a 22 y/o college graduate with no real life experience and no work experience counseling your children, teaching them life and death choices? Heck no…so why should they be placed in a position they are not qualified for that involves the protection of or destruction of families? What kind of madness would allow such people to make such decisions with regard to the lives and well-being of others? An insane and unchecked one….

The second part of the answer is covered in the next few paragraphs. For over 30 years, there have been numerous case file studies and for over 30 years there have been no designated standards, no outlined steps and procedures that are consistent,

and the whole process has been and continues to be based upon subjective and secretive court procedures that require only the willingness to stretch the truth to destroy a family.

In 1983, Theodore Stein and Tina Rzepnicki examined the literature on decision-making by child welfare workers, explaining: "For over two decades, researchers have made efforts to discern the processes that child welfare workers engage in when making decisions for children. This research proceeded on the assumption that if key decision-making points could be identified and the decision-making process described, a decision-making framework to guide child welfare staff in making critical choices could be developed inductively from practice knowledge. Unfortunately, consistent decision-making principles have not been identified," they concluded

In 1994, Duncan Lindsey, Associate Professor of Public Policy and Social Research at the University of California, again reviewed the literature, concluding that "the child welfare field does not possess an adequate scientific knowledge base for determining which cases are best served in-home and which need out-of-home care."

In a follow-up study, Phillips, Haring and Shyne sought to assess the reliability of the decisions made by experienced child welfare caseworkers and judges in recommending for or against removal. In examining 127 child placement cases, the researchers found considerable disagreement between not only judges and caseworkers, but between judges themselves. The overall agreement between the six judges examined was less than 25 percent. Of particular significance was the wide variation between two judges, with one being four times more likely to recommend in-home services than another is. While the one judge acceded to removal in 17 percent of cases, the other acceded to remove the child in an astounding 72 percent of the cases studied. Even in those cases in which the judges agreed to remove a child, they varied substantially on the type of case plan or services to be provided

A 1995 South Carolina audit notes that while one county office confirmed abuse or neglect in 89 percent of reported abuse and neglect allegations, another county office confirmed only 14 percent. The variations in substantiation rates were found to be even wider when the broader category of "maltreatment" was alleged. The variations in substantiation rates varied from one county office's determination that no maltreatment allegations were unfounded to another county office's determination that 74 percent of the allegations were unfounded. The auditors noted that "the percentage of confirmed allegations among counties varied more than could be explained by outside factors."

Theodore Stein analyzed 68 intake and service worker interviews, finding that between 59 to 78 percent of the information gathered during intake the interview was not related to the objective of reaching a placement decision, and between 22 to 90 percent of the information in service interviews did not relate to the interview goals set by the workers.

Add to this the lack of clear definitions as to what constitutes child abuse and neglect, coupled with lack of clearly defined intake guidelines, and it becomes clear why Child Protective Services cannot deliver on its mandate to protect children--even as it continues to needlessly destroy families. As Douglas Besharov explains: "These laws set no limits on intervention and provide no guidance for decision making. They are a prime reason for the system's inability to protect obviously endangered children even as it intervenes in family life on a massive scale."

As Jeanne Giovannoni and Rosina Becerra explain, "Many assume that, since child abuse and neglect are against the law, somewhere there are statutes that make clear distinctions between what is and what is not child abuse. However, this is not the case. Nowhere are there clear-cut definitions of what is encompassed by the terms."

Existing legislation invests judges and state agency personnel "with almost limitless discretion to act in accord with their own child-rearing preferences in areas generally under the exclusive control of parents." Identifying children whose placement should be called into question requires more than vague and subjective language, and statutes must be revised in order to "prevent judges, lawyers, social workers, and others from imposing their personal views upon unwilling parents."

Add to this mix a darker side of human nature in the form of overt bias or prejudice. A recent grand jury investigation in California examining the disproportionate impact of foster care on black children suggests that racial bias may indeed be a significant placement factor. The jury reported that case records and court reports for white children were "consistently more detailed, better prepared and oriented toward family reunification, adoption or guardianship" than cases involving minority children. While white children's files had "well-documented" plans for permanent placement, such as adoption or guardianship, minority children's files did not contain any evidence of a permanent placement plan.

Then we must also recognized and emphasize that there are overwhelming financial incentives paid by the state and federal governments, there has existed and is now a legalized-illegal kick-back system that pays the judges from the counties CPS funds (conflict of interest and bribery by legal definition), case workers bonuses, attorney bonuses and fees, and CPS-DFCS agencies BONUSES for every child put through adoption or kept on their books. There is no clear method of decision making, no clearly defined rules, no clear legal definitions, and there is no protection for children and families from this system.

What do exist in this system are blatant torte violations, overt bribery and payment to judges who gain larger financial benefit by supporting continuation of cases and removal of children. This is interwoven with two systems of training, one sanctioned and explanatory of the rules and requirement of the law and the other done "informally" that instructs CPS-DFCS workers how work around the other one.

They are untrained and inadequately educated to perform the job, but are quick to "informally" be taught how to intimidate and coerce parents into cooperation, to generate false allegations and present them in a "closed-door courtroom" that allows no representation of the parents to be present and requires nothing more than their false testimony, perjury, to take your children.

Chapter Four
The Process used to Kid-jack Our Children

I will get around to given you actual cases as they have been given to me in Chapter Seven, and I will present both the most horrific and most surreal examples of what CPS-DFCS has been up to in Chapter Eight. However, you need to understand how the unlawful removal of an innocent families children is performed and how they get by with it thousands of times a day across the United States.

One method that is common around the country is rubber stamping of orders without review of the orders or evidence by an actual judge as Heather Catallo of ABC in Detroit reported, "When the judge dismissed the criminal charges against Maryanne Godboldo – he called the child removal order that sparked the case unconstitutional. As the Action News Investigators first showed you – the order to remove Godboldo's child from her home was never actually reviewed by a judge.

This process is called rubber-stamping – where probation officers - NOT judges -- are literally stamping a judge's signature onto the orders used to take children away from their parents. That means in Wayne County -- a judge is not looking at the evidence in a case before a child is removed – and court experts say that's illegal – and it has got to stop. It is ridiculous to go in to remove, in this court's opinion, somebody's child based on this order," said 36th District Court Judge Ronald Giles.

"This is testimony from Godboldo's civil custody hearing that child advocates call shocking:
Question: Are any of those probation officers attorneys?
The supervisor answered: No.
Question: Are any of those probation officers sworn referees?
Answer: No."

The judge who is supposed to be reviewing these cases has not been removed from the bench nor have any charges been filed against the judge or the probation officers who violated their oaths of office and broke federal laws. I assure you that if any of the public outside the juvenile court system acted in this way, we would be disbarred, fined, charged criminally, and forbidden to practice law.

When there is an actual "hearing" it is not a legal hearing at all. The defendant or accused is not allowed in the courtroom, nor their attorney, nor any representative of the accused. The accused is not allowed to face their accuser, to hear evidence against them, or to enter testimony or witnesses into the process. Often the actual transcript of what the caseworker, the agency attorney, and the attorney for the child(ren) is never made available to the accused or their legal representative. Instead, they are given a summary of the judgment against them without seeing their accuser, what the actual testimony was, or that they have only 30 days to refute and file a counter-claim against the CPS agency and its representatives.

There are very important differences between all other court procedures and those installed inside the mafia style system known as the "Family Courts". In any other process brought before the court, you are given representation, the right to face your accuser, the right to present witnesses, and the right to actively participate and be represented during ALL phases. In "Family Court" these rights are violated in every case because that is the procedure. The way in which they operate is to intentionally deny the rights to fair trial.

This initial hearing is rife with fraud and perjury. The caseworker and attorney for the agency know exactly what needs to be said to the judge to get a judgment. The records are going to be sealed and are often altered six months to a year after the fact to reflect "new accusations" that CPS caseworkers have created to keep the case in the system. I wish with the most sincerity that this was not the case. However, to be honest and forthright, I have to say that this is the norm in caseworker testimony not the exception.

It should be clearly stated that the judge is going to financially gain from taking your child as well. Therefore, the "sell" is often a matter of formality because they were going to take or keep your child in the system for profit regardless of the facts of the case.

They are protected from prosecution and have absolutely no fear of being charged within their system. The records are nearly impossible to get for use in federal court to file complaints and judges have been observed on the record allowing caseworkers who are under scrutiny to refuse to answer questions of the parents lawyer if they "feel uncomfortable answering anything" (Klamath County, OR). This creates a different situation on the stand, while everyone else would have to either answer the question or plead the fifth to prevent from incriminating ourselves; caseworkers are being given a pass when their conduct has violated the law.

The other reason that caseworkers and CPS supervisors have so little fear with regard to giving false testimony, committing perjury, and filing fraudulent paperwork with the "Family Court" system is because they cannot be pursued after 30 days nor are the parents and their attorneys allowed to protest, contend, or demand corrective actions of that system after 30 days.

They can lie, enter it into record (perjury), and file the papers with the court (fraud), abduct your child unlawfully, without cause or exigent reasons and after 30 days there is not a damn thing an innocent family can do to get justice or criminal charges proffered against what have been active criminals for decades in many cases.

For any other citizen of the United States we have no such protections. We will be charged and brought to trial for up to two years after the discovery of our fraud or perjury. In the case of law enforcement officers, every case they have been witness to comes under scrutiny when they are found to have committed perjury or altered evidence in a single case. Not so with CPS caseworkers. They are not held to discovery of their crime, but are held only to 30 days after the initial hearing, during which the parents had no representation.

The next trick of the "Family Court" system and CPS agencies is to schedule the next hearing to happen after the 30 day limit. If you have a court appointed attorney, I assure you that they will not demand a sooner date, nor in any of the over 800 cases that I have read through did any court appointed attorney even attempt to defend innocent parents against the false claims asserted by the CPS caseworker.

Several CPS workers in the state of Oregon have reported to me that this process is acted out in this way for the express purpose of preventing the parents from being able to contest the errant testimony of the caseworker. Several others from other states have said this is an unprotested matter on the part of "court appointed attorneys" as well because they are going to go along with the system against the best interests of their clients.

The important fact about that is that most of the "court appointed" attorneys rotate their representation and are assigned to cases by someone within the system. These attorneys may represent the CPS-DFCS, a caseworker, a child, the court, or the parents…or any number of them on the same day. This sets up an instant conflict of interest because to act against CPS or the courts is to risk being disbarred or banned from practice within the system.

Something else to be aware of in order to understand how they have and still continue to act against the law, commit fraud and perjury, and unlawfully kidnap children is best said by a lawyer for a family in Texas, "DFPS in Texas is allowed to govern itself - all the Administrative Reviews are performed by a DFPS agent at a DFPS location. There is no outside governance of CPS in Texas."

This is the case in nearly all of the states and where there is outside governance allowed and planned for they are panels which have no real power or authority to change the system or to protect families. Nowhere in the United States have these "panels" or board attempted, in any way, to address the atrocities and illegal actions of CPS-DFCS. Nothing has been done federally even though this problem is known to be a nationwide corruption.

An attorney currently establishing a class action suit in Florida and Georgia has told me that in 500 cases his staff has reviewed, there also was no attempt by the court appointed attorneys to act before the 30 day limit though they are 100% aware of the exceptional statute of limitation allowed for CPS workers who have committed fraud.

CPS caseworkers from several other states have substantiated both my research and the claims of the four major attorneys working to establish class action suits in Florida, Georgia, Oregon, and California at this time. Court appointed attorneys are more often seen working with CPS caseworkers rather than acting on behalf of the clients they were and are required by law to represent.

A CPS caseworker who resigned recently from the offices in Eugene, OR reported that because the system is so closed off from all other branches of government, the attorneys are often representing CPS agencies and caseworkers in the morning while feigning representation of the children and parent's best interests in the afternoon. This is absolute conflict of interest in ways deeper than most and many other caseworkers have verified this to be the norm nationally.

In any other instance an attorney can not represent an individual or agency and the defendant who will have to face that agency and caseworker the same week. I have had attorneys tout their requirements under the law only to find out that they have unlawful conversations with CPS caseworkers outside of the courtroom, without the consent of the client, and reach agreements that are not representative of the client. All this occurs fully without the client's knowledge of such a relationship.

An attorney representing families in Michigan reported cancellation of his other assigned clients by the firm for which he worked when he decided to help over two dozen families fight CPS agencies who entered false testimony and fraudulent claims into the court in order to remove the children from their homes. One of the firm's major partners was a contracted attorney for the

CPS agency who promised that he would never find work in the county again if he pursued the case. Attorneys and caseworkers who have stood up for parents and children against the court system and CPS agencies have found themselves arrested, fired, harassed, and have had their own children removed from their home under false circumstance. In many instances attorneys who have taken cases have been disbarred, defamed, and slandered by those inside the system.

In 2003, Barbara Johnson was disbarred by the Massachusetts Board of Overseers because of the Class Action she was working on against CPS and Family Court.

Attorney Gene W. was disbarred when he filed a complaint in federal court. The tapes of his clients hearing in "Family Court" as well as the transcripts had been edited and falsified testimony was found in the edited versions.

Dr. Steven Baskerville states that is it is "surprisingly easy for these agencies to remove children, alter court records, and enter false testimony."

Dr. Charles Heckman points out that there is "no penalty for a government lawyer who incite their witness to commit perjury".

John Wolfgram is now a "blacklisted" attorney for standing up to the system who states that "there is complete immunity" for those who are committing this abuse.

Elsa Baumgartner was disbarred for doing her job, defending her clients and the constitutional rights afforded them. Her crime? Doing it inside a system that has no respect for either and reporting case fixing, altering of court records.

Many attorneys who specialize in racketeering have called the system the "greatest racketeering scheme in the history of the United States". One states emphatically that when these agencies demand that you attend "classes" or "educational" programs or they will take your child, it is "racketeering by definition of law".

Ed Truncelito filed a complaint with his state bar stating "racketeering"…he was disbarred for doing his job and defending the legal and justice system.

The judges within this system have given themselves "Absolute Immunity" under the term, "sovereign immunity" which once stated that "malicious or corrupt" was an exception. Having gotten by with that, the words "malicious and corrupt" were removed. With no protest, the attorneys within this system were then given "qualified immunity". There is no basis for this anywhere in the legal documentation of the nation, and is expressly forbidden by constitutional and judicial statutes.

This is both unconstitutional and a violation of every law ever written, every statute ever written to effect a "checks and balances" process within the judicial system. This should also clearly explain why they are able to act against the laws, statutes, and constitutional provisions with impunity.

The statutes, legislation, and constitution clearly states that "all men are equal before the law"...well, all except those who have a system designed, run, and manipulated to protect them from criminal prosecution for the crimes they have committed for over 25 years.

Many attorneys and activists working against the judicial system and these changes are quick to point out the words of Thomas Jefferson, who was himself an attorney:
"The germ of destruction of our nation is in the power of the judiciary, an irresponsible body-working like gravity by night and day, gaining a little today and a little tomorrow, and advancing its noiseless step like a thief over the field of jurisdiction, until all shall render powerless the checks of one branch over the other and (the judiciary) will become as venal and oppressive as the government from which we separated."

History likes to repeat itself. Wise men of their times saw tyranny and oppression and recognized the source and means being used. Thomas Jefferson and many others knew what powers men would seek to gain and the danger of letting them do so without conscience or consequence.

Today, we find those words and warnings alive and well inside the judiciary in powers tyrannically wielded by the "Family Courts" and the CPS-DFCS agencies that they protect.

Chapter Five
What about the Rules?

There are in fact rules of conduct and procedure that are written and included in guidebooks, learning tools, and course work that are offered to CPS-DFCS workers, the attorneys, and all those inside these secret courts.

However, the very fact that these are "secret" or "secretive" courts with "ex parte" hearings that do not allow the accused or their attorneys to attend should be the first and only thing I need to say to completely invalidate the whole system. This alone is a violation of every accepted means of conducting a hearing, allowing witnesses, allowing the accused to face their accuser, and due process of law.

The judge then issues a "civil orders" without you or your representatives being present. This is done in violation of your rights because you were not there and in violation of judicial procedures, the constitution, and every legal statute regulating this process.

There is "sovereign immunity" for the judges even in the case of "malicious or corrupt" behavior. Statutes have been rushed through state legislatures by the state bars to allow for "legalized bribery of judges through kick-back from the county (CPS-DFCS agencies) that illegally launders federal money to the judges as "bonuses".

The attorneys are also given "immunity" which means that they can act against their client in cooperation with CPS-DFCS and the courts to commit fraud, perjury, and violation of constitutional rights and you can do nothing to prosecute them.

I want to note here that any statute put in place by a court system or legislation passed by a state or federal body that is in violation of the constitution is an "Invalid" law or statute. They can try to hide behind immunity, but soon the federal courts are going to be filled with complaints based upon the unconstitutionality of their proceeding, immunity, and illegal actions. I personally will be heading one such complaint as the 1st party of record.

As part of the preparation for this book, I had attorneys and paralegals assist parents in writing a complaint to the bar association. I myself drafted many of the letters and sent second letters of complaint with the discovery papers showing the extent of the perjury, fraud, extortion, and coercion. The bar associations have refused to hear cases of misrepresentation and acting against their client's interests in ALL FIFTY states.

Some were part of my grouping and others are cases over the last five years refused to be heard by the bar association though there are obvious conflict of interest, "Bar" rule violations, and procedural violations on top of improper representation of their clients. I point out again the "immunity" that those working inside this system have been given.

In addition to the "Bar" associations, letters have been sent to the State Attorneys General, to which the response was the same. Often letters stated they "had no authority" to act on behalf of families who have been defrauded, extorted, misrepresented, perjured against, and have had their children unlawfully taken by the "Family Court" system.

In two instances in which I was personally involved in, I assisted with the drafting of the letters, articles of complaint and discovery, and request for remedy that were sent to the Attorneys General. In Oregon specifically, I was able to find out that the Attorney General's office contacted the district supervisor and "demanded" that this problem be resolved. So much for not having influence.

However, I want to point out that both the Bar Associations and the Attorneys General are correct in their answer which was the point in writing them in the first place. "Family Court" and CPS-DFCS has sovereignty of its own. There is no oversight board that can issue warrants and arrest those breaking the law.

No state agent can enforce state or federal law upon those inside the system because they are immune from the rules, immune from prosecution, and exist outside of the rules and laws of the states. They are protected by the "immunity" they have given themselves. Meanwhile, the federal government, federal courts systems, and Department of Justice are doing nothing to bring CPS-DFCS agencies or the "Family Courts" any semblance of accountability for the laws and constitutional provisions that they break continually.

Though the FBI and Department of Justice could pursue Racketeering charges against all participants in the system, they have as yet refused to do so. This falls under the "rules" that most people ask me about.

Racketeering by definition is a person or entity who commits crimes such as extortion, bribery, loansharking, and obstruction of justice in furtherance of illegal business activities.

According to the "Racketeer Influenced and Corrupt Organizations Act" the CPS-DFCS agencies and "Family Courts" qualify for prosecution as co-conspirators in a "racketeering organization". Under the law, the "Rules", the meaning of racketeering activity is defined in "18 U.S.C.

This includes any violation of STATE STATUTES against gambling, murder, KIDNAPPING, EXTORTION, arson, robbery, BRIBERY, counterfeiting, THEFT, EMBEZZLEMENT, FRAUD, OBSTRUCTION OF JUSTICE, MONEY LAUNDERING, etc...

Please note that all those listed above that I placed in all caps are crimes committed by CPS-DFCS and "Family Courts" in every county of the United States on a daily basis as part of their procedures and normal way of doing business. That is the fact, no exaggeration needed.

The rules also clearly state that no child is to be taken into "state custody" or "place in foster care" if a suitable family member or close family friend is available to take the child. However, this is violated daily and as part of the process without any consideration what-so-ever.

Chapter Six

The Financial Incentives

The corruption and violation of constitutional rights goes without saying. It is not one office in one state in one county. This is the case in every state, every county, and every office. There are NO exceptions that I have found in six months of research, nor have any others involved in fighting this problem found an exception.

Some would like me to acknowledge that there are those within the system who are there to do what is right for families and children. Others and I will admit that many enter into services with CPS-DFCS and the "Family Court" to make a difference.

However, just as Eichmann had the opportunity to walk away from his post knowing the atrocities that were occurring at the end of the rail-road lines and was executed for his participation in the crimes against men and humanity, anyone who is part of the system knows what the system really does, how it breaks the laws in order to profit from destroying families and are just as guilty as those who commit the crimes in remaining silent and actively helping them to do so.

Eichmann never killed a single person; many CPS-DFCS workers never break the law directly or violate the constitutional rights of children and parents, but in signing the papers, processing the orders, and allowing it to continue they are just as guilty, just as complicit, and just as criminal in their actions and inaction.

The financial gains made by "lawful-illegally" kidnapping our children are immense. For every child taken into the system 25 people derive their paychecks. The industry set up around this system is known to make an average of $85, 000 and up to $250,000 for every child taken and the case being held for six months.

The industries set up around CPS-DFCS and the "Family Court" system make tens of billions each year providing services "required" by the "plans" initiated through the institutions.

It is not coincidence that counselors of one service provider may work for CPS, the court, the attorneys, and in multiply provider agencies. It is also not a coincidence that attorneys, judges, and high ranking workers within CPS-DFCS agencies set on boards or have paid "advisory" jobs with those agencies they contract to do the services required.

Graft is best described as unscrupulous political corruption in which a political party or government employee uses their position for personal gain. Graft also happens to be another of those "racketeering" issues and is another common element within the CPS-DFCS and "Family Court" system. Those within the system contract and direct parents, children into the services companies that they own or from which they receive pay or financial gain.

Nev Moore, Founder of Justice for Families and many other have outlined the financial incentives for unlawfully taking children from their home…"Child protection is one of the biggest businesses in the country. We spend *$12 billion* a year on it. The money goes to tens of thousands of state employees, collateral professionals such as lawyers, court personnel, court investigators, evaluators and guardians, judges, and CPS contracted vendors such as counselors, therapists, more evaluators, junk psychologists, residential facilities, foster parents, adoptive parents, MSPCC, Big Brothers/Big Sisters, YMCA, etc."

Service providers, called vendors, hold multi-million dollar contracts with CPS. Families are ordered to engage in "services" with these contracted vendors, all of whom bill Medicaid at enormously inflated rates for "services" that are often inappropriate, unnecessary, and completely irrelevant to the families situation, or to child maltreatment.
Judges and CPS-DFCS workers are allowed to set on the board, work as advisors, and to be intricately married financially to these "service providers". The financial benefits received in doing so are protected and no charges of malfeasance have ever been proffered. Yet, state and federal laws specifically prohibit this practice for obvious reasons of corruption, malicious judicial decisions for profit, and benefits that create built in conflict of interest.

Families are coerced into participating by intimidation and the threat of losing their children. In return for fraudulently collected profits, the contracted vendors provide manufactured "evidence" of child maltreatment to support the claim that they need to keep the child in foster care. In return for this they receive more federal funding. Now this is truly job security."

In California, Richard Fine was the poster child for an attorney doing his job as it was intended to be performed. He was a chief attorney in the prosecution of anti-trust violations in Washington, D.C., and was respected for his moral and ethical standards. However, when he exposed the "kick-back" system in place in LA County, CA he became a political prisoner.

He was held in jail without charges, without trial, without an attorney, in solitary confinement, and without any hearing for over a year by order of a judge who was not only guilty of the crimes Fine was seeking relief from. The judge that was able to do this was listed on the federal complaint.

The county made payments from Federal funds to the judges and the judges in turn made decisions that favored the county financially. The diversion of the funds is laundering, and the payment to the judges is open-ended bribery. Both illegal, both specifically addressed in the constitution of California, and both still illegal regardless of the rushed legislation that was passed illegally by the California Senate.

In the case of CPS-DFCS and the "Family Courts", this was only good for the county if children were removed from their families or were mandated into the CPS-DFCS system. This of course, included all the mandatory uses of the "service providers" who also benefited from the arrangement. In the case of the criminal courts, longer sentences benefited both the county and the private industries that owned prisons. This behavior is unethical, illegal, and immoral without a doubt.

In response to the discovery and case filed by Mr. Fine, the judges and legal lobby of California went to the state legislature and had statutes written that gave the judges not only immunity for all the bribes that they were taking, but immunity from prosecution for the bribes they were taking for a decade prior to the bill passing. Immunity for judges who have been in violation of torte rules, the California constitution, and federal laws for over a decade. So much for justice.

Seeing that there were "legal limits" clearly defined in state statutes, the Bar Associations and legal lobbyists have succeeded or are in the process of getting similar legislation passed in every state in as quiet a manner as possible.

If this were illegal, Graft and Racketeering for decades, unlawful in every state and according to Federal Laws, how can a legislature not only make it legal, but given immunity from prosecution to the judges and county officials who had been doing this for over a decade? The answer is that they cannot and any legislation doing so is unconstitutional and a violation of federal law not to mention the constitution of California.

I seem to remember that "no law can violate the constitution; the constitution is the supreme law of the land". However, since most of the population of the United States has no idea that this is the case they have no idea that their legislature, judges, courts, and attorneys are violating the constitution on a daily basis.

You need understand that this is happening throughout the judicial system not just in "Family Courts". Judges in Luzerne County Pennsylvania were being paid by privately owned prisons to send innocent people to jail for "holding" or "hold-over" so that the prison could bill the county/state for housing the innocent people. This is fact. They are being charged only after they decided 6000 cases and at least 2.9 million dollars was paid to them by the prisons.

Now wait a second. They took kick-backs adding up to 1.2 million over five years and were paid to make decisions in favor of the county and the private businesses. Same kind of issue we see in the "Family Courts" and CPS-DFCS agencies, so why are they not in jail awaiting trial?

Judges within the "Family Court" system get legalized "kick-back" from the county which is according to legal definition, laundering federal money designated for CPS-DFCS family programs to do so.

There is no money to be made by returning the children to the parents and closing the case. The "permanency plans" are far from it. Every six months for as long as they can get by with it, the casework will change the plan and coerce the family into signing the new plan under threat of losing their child. The longer they can keep the parent and child in the system the more money they make. Often, they have the parent take classes offered by "service providers" more than once.

According to an Oregon caseworker who resigned, "anyone who suggested anything that went against the planned adoption procedure had their testimony and opinions deleted from the records...the parents were run through the ringer and forced to jump through every hoop, but the case supervisor was never going to allow anything but adoption and had said so in the office from the beginning. That was the only way she and the county could get the bonus money. When I protested, I was ostracized in the office. She (supervisor) even said that it would be easy to take (my) children from a single mother."

Another worker in the CPS offices of Oregon stated, "There is never any intent of reuniting with the family if at all possible. I was so disgusted that I quit, took a smaller salary, and will never speak of the supervisor in our office again. When she found out a caseworker had made up evidence in order to take a child, she said 'no one will ever know' and the girl got an extra day off. 'That's money in the bank', I heard her say many times over several months...she and the others who went into court were cold and calculating, always thinking about the minimum charges they could make to get the kids out of their homes."

The same worker stated that a state conference actually had a four hour presentation on how to get the parents to consent to a "permanency plan" without cause to avoid court hearings.

Several other caseworkers in the state of Oregon have reported to me and numerous reports from nearly every other state show that caseworkers are ostracized for wanting to reunite the families too quickly even though there is no cause found to have removed the children. Many personally told me that they have been threatened with discharge if they could not find a way to get their "removal" numbers up.

The payments vary, but the reason for adoption, even if only to another family member is the amount of money that the CPS-DFCS agency and the county make in doing so.

The process is specifically aimed at poor and low-income families. CPS-DFCS will make allegations and coerce the families into "permanency plans" that require the families to participate in the "service provider" programs. Here we have money for CPS-DFCS and the providers.

However, since the funding to "assist" these families is being diverted to illegal bribes and bonuses for attorneys, CPS-DFCS workers, and judges there is little funding supplied for payment for these services. Knowing that the families cannot pay, they are ordered to "meet requirements" that from the beginning is impossible for them to afford.

The intent is that when they are unable, the child will be put up for adoption or placed in the foster care system that makes even more money for the CPS-DFCS agencies. It is important to know that in many offices, adoption is the ONLY option ever truly on the table regardless of the manipulation and profits made by suggesting otherwise to the parents. The family will be fortunate if one of their relatives can qualify and even if a family member gets the child(ren) the CPS-DFCS caseworkers are trained to get the adopting persons to enter into programs that essentially guarantee monthly or annual payments to that office until the child is 18 years of age.

The amount that CPS-DFCS receives varies, but CPS-DFCS officials who spoke to me gave me dollar amounts that are staggering. An initial payment of $4,000 to $12,000 is paid out when the adoption is processed. In many instances, with proper manipulation of the "adopting" parents, there are programs that will pay the CPS-DFCS office tens of thousands more up to the time the child is 18 years old. Yeah, I said it twice. Why? Because it is all about the money.

Many counties and states have now implemented "Fast Track" adoptions through CPS-DFCS agencies and the "Family Court" systems in order to more quickly get access to the Federal Reimbursement program. This is money in the millions of dollars and does not include the payments to the service providers, court, the judge, the caseworkers, and the attorneys.

The profit is in taking children, keeping them in the system as long as possible, and if at all possible getting an adoption to someone, family or not. Many high ranking officials will point out that this "adoption bonus" does not cover their costs. However, they do not mention all the other compensatory packages that all the parties involved receive nor will they even address all the funding diverted to pay for services in the cases of parents falsely accused and coerced into the programs offered, demanded, by CPS-DFCS workers.

The other thing that none will address is the billions in dollars that is rolled out to the foster care system. Unlike the beliefs of most, the reality is that the foster parents receive less than ¼ of the funds the CPS-DFCS offices are paid while the child is in foster care. The amounts of money made by unlawful removal of children are staggering, many layered, and unconscionable to a moral and ethical person.

Chapter Seven
Whom Can You Complain To?

I have numerous organizations listed and described in this chapter and I would suggest that you write them all. Send a letter every day. Have your relatives send a letter every day. The post office offers bulk rate for mass mailings. Send all of them at the same time. If you get an answer that says they cannot be of help, send five letters a day from you, your relatives, and your friends.

Send emails to every news outlet in your region, state, and nationally. Send them every day and send them even when you aren't sure if they will help you. Send email and letters to the White House, your Senators and Congressmen, State Representatives, your State Attorney General, your District Attorney, the Bar Association. SEND THEM, SEND THEM, AND SEND THEM.

Find the largest churches in your state and region and make sure that everyone listed in their prayer groups, the pastors, elders, and everyone whose email is available gets one from you outlining the real situation. Send them letters with the rest.

You have to make your case and unfortunately are going to have to flood mailboxes for weeks before you get a reply and months before you get action.

Join every "ANTI-CPS/DFCS" group you can find online and in your state. Sign every one of their petitions and actively start your own in your community. Be an active member by sharing your story and arranging protests at the courts and CPS-DFCS offices.

Research everything you can find on CPS-DFCS statutes, practices, procedures, and then forget it. They don't follow their published rules, the law, or the requirements of court procedure. They don't adhere to constitutional protections. They don't tell the truth and don't serve "Children and Families".

With that in mind, you then find out what they are actually doing, to whom they have been doing it to, and what others have been going through. You find the resources that will give you the best information to fight the system as it is now and you spread it online, in bulletins, and in letters to everyone you know.

You find the videos on YouTube, news reports that were actually made, testimonials, statements, and cases that show what you are saying about the fraud and false practices of the "Family Courts" and CPS-DFCS are real and the dominant process within them. Then you post them on your webpage, your Facebook, MySpace, and any other you can find. Make flyers and disperse them at festivals and events so that others can no longer ignore the corruption and abuse of our children, our grand-children, and the children of our friends by the agency created to protect them.

The primary motivator behind these illegal actions and profound violation of law, constitution, and rules of conduct is money. You have to make it unprofitable for the "Family Court", CPS-DFCS, the attorneys, and county to continue to deny you your rights and to violate your child's right to be with you for profit.

The best way to do that is to take away their money.

In the case of the attorneys, (I point this out to those who are going through this book as victims of the system, who know victims of the system, or who had no idea how corrupt the system is). This is the method and means to get these attorneys out of the courtrooms and to end their financial gains at the expense of children and families. I would suggest again that all groups who are bound together by this issue, bring class action suits against the bar association in your state for not disbarring attorneys inside the "Family Court" system for violation of the constitution of the state and the United States. That information is already in this book in a few places. As for your individual attorneys who are not adhering to the law and oaths they take, name individually every attorney who failed to represent you.

The county can also be held accountable for allowing CPS-DFCS to violate your rights and act against the constitutions. In order to do so there are forms available in every state and every county for reimbursement of loss. They do not make the forms public knowledge and I have yet to find any state or county that has them online for you to obtain, but they are required by law to provide one to you upon request.

You go to the County Court House and get the form. You enter "CPS Fraud and Perjury" as the cause and explain it in the space on the form. You should include racketeering, kidnapping, bribery, coercion, extortion, and willful disobedience of state, county, federal laws.

You can claim all lost wages, any medical or medication expenses, any expenses they have accrued to you such as their "parental support fees" while your child is in foster care, pain and suffering, mental anguish, and I have seen many include a "per diem" fee for every day that the child has been separated from the family.

If everyone who has had their child unlawfully taken from their family by CPS files a form in the next week...there would be millions of claim forms entering the system. By the way, fraud and perjury by a "county representative or official" is acceptable and reasonable cause for reimbursement by the county and state.

If they deny your claim, you have the right to appeal in front of both the county and state. You should immediately file the appeal and request a hearing. Your burden of proof is less than in a criminal court. Your paperwork and case history will be sufficient to show you were both defrauded by the county and that it was the result of illegal action by the CPS-DFCS offices and the "Family Court". Again, I want to point out that regardless of what legislatures and county councils pass, if it is in violation of constitutional law then it is not a legal defense. You need to point this out when you fill out the paperwork and during your hearing.

Get everyone who you know in your support groups and "Anti-CPS/DFCS" groups to do this. The more pressure you put on the agencies, the sooner things will change. Additionally, I should not need to point out that they do owe you much more than financial compensation for the actions of those under their authority, they owe you justice as well.

If they want to illegally take your child, file false statements and commit perjury in court, and work within a closed, illegal, and unconstitutional court system, then you should work within the one you have been provided. Take every dime the county has made by kidnapping, extortion, and fraud back into your pockets. If you don't want it, give it to charity or one of the organizations working to put the criminals inside this system in prison.

Racketeering is still covered under federal law. File racketeering complaints with your local, state, and federal Attorneys General. State the initial fraud committed by the CPS agency. State clearly the reasons for it being fraud.

Include the fact that they are using your child to unlawfully "Extort" and "Coerce" demands from you and your family for financial gain. Then point out that the funds diverted to judges and CPS workers as bonuses is being taken from funds intended to be used to help children and families. This is laundering and bribery. THAT is racketeering. IT is ILLEGAL. IN ADDITION, you have the right to be represented by your prosecuting attorneys.

The next step is to write your state bar association. Point out that your court appointed attorney has worked in cooperation with the CPS to violate your rights, has failed to inform you of your rights, and had not represented you in your case.

This information should and must also be sent to your county prosecutor, state Attorney General, State and Federal Senators, Congressmen, and any media you wish to inform of these unlawful actions.

There should also be matching documentation sent to your STATE assemblymen who oversee the state's department of health and to every committee member of the US Senate that oversees the Department of Health and Human Services...I am including their names below. When this changes the list will not be accurate, but now you know what committee to write to and they will be listed on the Senate website…

This is the complete list of current members of the Federal Committee overseeing the Department of Health and Human Services. CALL THEM, WRITE THEM, SEND THEM THE COMPLETE COMPLAINTS you have including NON-representation by court system appointed attorneys. Then get every friend you have and every family member you have to do the same.

If YOU WANT your children and others children protected from this scam? THEN START WRITING TODAY....

Democrats by Rank
Tom Harkin (IA)
Barbara A. Mikulski (MD)
Jeff Bingaman (NM)
Patty Murray (WA)
Bernard Sanders (I) (VT)
Robert P. Casey, Jr. (PA)
Kay R. Hagan (NC)
Jeff Merkley (OR)

Al Franken (MN)
Michael F. Bennet (CO)
Sheldon Whitehouse (RI)
Richard Blumenthal (CT)

Republicans by Rank
Michael B. Enzi (WY)
Lamar Alexander (TN)
Richard Burr (NC)
Johnny Isakson (GA)
Rand Paul (KY)
Orrin G. Hatch (UT)
John McCain (AZ)
Pat Roberts (KS)
Lisa Murkowski (AK)
Mark Kirk (IL)

Chapter Eight
The CASES

Many who read through this chapter are going to find their mind and morality demanding that they believe that these are "rare" or that those things I am including here are the "exception" not the normal condition of the CPS-DFCS and "Family Court". I truly wish I could tell you that it is the case, and confess that when I was first approached on this issue, I wanted to believe that there was "no way that this can be happening".

I also wanted to believe that the Attorney's General, the Bar Association, the U.S. Department of Justice, and any number of others would never allow such an atrocity to take place continually in my country for over 20 years. I was wrong.

In truth, as I have done the research I have continually said "there can be nothing worse than what I just found" only to find things much, much worse. I confess openly that I have more than once had tears of pain, anger, and absolute disgust pouring from my eyes as I have read, watched, and researched the corruption and abuse that is the normal procedure within CPS-DFCS and the "Family Courts".
Heinous, abhorrent, despicable, illegal, unconstitutional, disgusting, horrific, evil…well, that is without using offensive language. The worst string of profanity would not even begin to cover the extent and depth of depravity, atrocities, and abuse of children and families by the agency and court created to protect, defend, and assist them.

I am going to recite briefs in this chapter and at the end there will be a list of websites, web-video, and new pages for you, the reader, to find the truth in what I report.

You need to understand that children are 600% more likely to die a horrific death in the custody of CPS and Foster Care than in the general population.

Case Number One: Adriana Cram taken into custody by Oregon CPS when her mother went to the DHS Services office seeking assistance with medical needs of her daughter. Adriana was then shipped outside the United States where she was murdered by being slowly beaten to death.

No parent or any other agency in custody of a child can move without notification to all parties and right to hearing on the part of the other parent. None of that occurred in this case and even with the death of the child, no one in the CPS system was charged with his or her crimes.

Case Number Two: Josiah David Carrol, Mollie Ruth Carroll, and Noah Adam Carrol died in custody of CPS and "no cause provided" was allowed to be listed, no charges filed.

Case Number Three: Christopher Dove died in custody of CPS and "no cause of death provided", an infant dead and no charges filed.

Case Number Four through almost NINE THOUSAND over the last 20 years: CHILD DIES in CPS CUSTODY, no charges filed and caseworkers not discharged.

Please note that these NINE Thousand are not by any means all of the children who have died in custody of CPS and the Foster Care program. This is just about 1/3 of the total deaths and with rare exception not even the individual adult on site at the time of death has been charged with a crime.

Now, let us start over with another issue:

Case Number One: Adamaris Carmona, 2 year old in Foster Care. Listed as a "runaway" on their books and no one questions it? 2 y/o runaway?

Case Number Two: In LA County alone, over 1,000 children in custody of DFCS are missing and no one has lost their job as a result of this fact.

Case Number Three: Ariana and Tyler Payne. Brandon Williams, a caseworker who was dating an abusive father she was supposed to be investigating -- and was promoted when found out. Think every Arizona child who has died this year while CPS was supposed to be watching.

The problem with Case Number Three is just the opposite of most. Instead of taking the children, protecting them, and keeping them from harm…the agency is so overwhelmed by "fraudulent caseloads kept on parents and children we shouldn't have in our system, that they can't protect the children who really need it," the one caseworker from Pima County who would speak to me said openly.

Case Number Four: Marchella Pierce died at hands of her mother while CPS did nothing to remove her from danger.
Two child welfare caseworkers have also been charged in the girl's death. These will be the first among the social workers in the country to be held criminally responsible for the death of a child on their watch

although thousands of deaths and disappearances occur each year.

Case Number Five: Social Worker in California promoted and training others after court finds her guilty of fraud and perjury. NO disciplinary actions were taken even though found guilty of crimes.
The agency director Michael Riley stated on record that "the agency can state categorically that if SSA social workers had perjured themselves or falsified or hidden any information, those social workers would have been terminated." Mr. Riley had to have approved her promotion and her training position.

The mother in the case that the DFCS worker lied, committed perjury in court, took a full six years to get her daughters back. This Social Service worker lied, cost this family five years and no measurable amount of misery, pain, anguish, and suffering, but was promoted and not charged with perjury?

A little factual inclusion:
Federal statistics say that there are approximately 3 million reports of suspected child abuse and neglect each year.

The U.S. Department of Health & Human Services documents 900,000 as "substantiated", leaving over 2 million families per year FALSELY ACCUSED. Of the "substantiated" cases, 68% do NOT involve child maltreatment, according to the federal authorities. This figure is escalating at an alarming rate each year as overzealous reporting and frivolous intervention by CPS spins out of control.

Let me break this down for you a little more accurately and clearly, out of 3 million reports 2,612,000 are not cases at all. That leaves only 380,000 that are legitimate cases according to the USDof H&H. However, we also must clarify that this is "as reported by the County Agencies throughout the country." The numbers don't mean a damn thing.

Those within the system who consented to be interviewed specifically stated that 80% of the cases handled nationally are cases that should not be in the system and that only 20% of the cases that should be brought and kept in the CPS-DFCS and "Family Court" system are, in fact, pursued.
I thought that information to be upside-down as well, but after nearly 70 conversations with CPS workers throughout the country this is in fact the average of their responses. Seventy, **70 caseworkers,** stated to me that they believe at least 80% of their cases should never have continued beyond initial investigation and that just as many that resulted in the removal of the child were done wrongfully and against CPS rules.

Those whom I spoke to from California, Georgia, and Oregon emphatically stated that if a family is poor…as one worker said, "they don't have a chance in hell with a court appointed attorney." Another stated that "poor families are going to get hell and then some from our office and the others in this state. Since the timber industry is gone, this is how my county makes up for the lost money."

The former L.A. Director of Child and Family services and several social workers from within that system stated that in "over 70% of the cases, the children should NEVER have been taken in the first place."

"In Orange County, CA) there are between 3300 and 5000 children taken each MONTH, those numbers are higher in San Diego County and Los Angeles County on average" supervisory judge in the Appellate court system confided to me. "These numbers vary a little and could be a plus or minus five or six percent, but it is safe to say that this amounts to tens of thousands each month in California alone."

"The simple truth is that the CPS (DFCS) agencies must obtain more and more children every month, every year to justify their budgets and to get budget increases. This is tantamount to requiring illegal actions on their part to stay in business," a retired Federal Judge from Eugene, Oregon stated. He went on to say that caseworkers have "come to me asking for advice. They are openly saying that even though it isn't official…their supervisors and district managers are implying quotas for both removal and adoption as well as the number of children they need to bring into the system each week, each month to keep their jobs. This is unconscionable."

The simple fact is that these children do not have to be put up for adoption though this is often one of the goals they would like to accomplish if possible. Having this many children in the system brings in trillions in revenue to the CPS-DFCS agencies and all the service industries served by the illegal actions of the courts and agencies.

I could go on for several books if I were to list all the cases that have been given to me with explicit consent to use in this book, and there will be a follow-up to this book with all the cases in a much-shortened form for anyone to see. However, I also want to point out that right at this moment, anyone who is willing to use a couple of search engines can find thousands of cases very quickly where Parents were perjured against by CPS-DFCS workers and children died because of the failure of CPS-DFCS workers to do their job.

I propose that they begin taking custody of the children in the 80% of legitimate cases that come across their desk and they would not have to rail-road innocent parents through illegal action, fraud, coercion, extortion, kidnapping, and perjury. It would be just as many and legitimate, moral, and ethical.

Nevertheless, I should point out that much of what CPS-DFCS workers do is part of a power-trip mentality that the system breeds in their training seminars and offices. "It isn't any fun to take children away from someone who is going to be in and out of our system for years, someone who can make our agency and service providers money for a decade or more," one supervisor was recorded speaking at a meeting in Oregon. "We are going to have them anyway, it is more fun to f—k with people who have no idea its coming and to see how pissed of you can get them."

Yep, I actually heard the recording and saw the video that will both be going to federal court very soon. I hope she is jailed, fired, and the attorneys who were laughing with her get disbarred and jailed for conspiracy to defraud the federal government and to kidnap children. Those are just a few of the charges that are being presented to the judge, with evidence of course.

In Josephine County, Oregon, a CPS caseworker was recorded telling a parent that "you'll go along with me or I will just come in here and take your kids all over again." Again, I will tell you I have seen the video and personally know the caseworker who was recorded. It will be a personal pleasure to see someone with so little regard for family sanctity and the law, removed from her position.

In Klamath County, Oregon, a local judge is now under investigation for refusing to make a CPS worker answer questions under oath. "If you are uncomfortable answering, you don't have to answer," was her answer to the parents' lawyer harsh questioning of a worker who entered false testimony (perjury again) against the parents and falsified documentation after the fact. This hasn't come out in the media yet, but it should have months ago. Any other person would have been ordered to answer the question or to plead the 5th, refuse to answer on the grounds of incriminating self. The actions of the judge are violation of the parents' right to fair trial and to face their accuser.

In Florida, CPS-DFCS workers across the state are subject to a class action suit. The attorneys preparing the case have overwhelming evidence of perjury, fraud, coercion, and falsification of evidence. Additionally, alteration of paperwork after the fact, intimidation, and threatening parents who refused to accommodate the agency in falsely filed cases.

Currently in Oregon and many other states, cases are being prepared that will actually pursue conspiracy to commit fraud, conspiracy to defraud the federal government, and racketeering against the attorneys within the CPS-DFCS and "Family Court" system.

In Oregon, the list of attorneys within the system who have been shown by their appointed clients to have failed to represent them, or to actually have worked with CPS against them is long and getting longer. The Bar Association will not pursue them. However, their actions are direct violations of federal and state law. Once charged and convicted, they can then be disbarred, which is one of the goals of the organization in this state. Then the civil suits will be brought to reimburse the families for the harm that these lawyers have done in service to the money made by "NOT" representing their clients appropriately. That they also work for CPS, caseworkers, and the courts in other cases is going to lead to very easily proven conflict of interest cases in most of the list I have seen so far.

I point this out to those who are going through this book as victims of the system, who know victims of the system, or who had no idea how corrupt the system is. This is the method and means to get these attorneys out of the courtrooms and to end their financial gains at the expense of children and families. I would suggest again that all groups who are bound together by this issue, bring class action suits against the bar association and name individually every attorney who failed to represent you.

If you want to find cases and causes, there are numerous webpages. "CPS Corruption", "Child Protection Services Fraud",

"Innocence Destroyed" is a 3-part video made by Bill Bowen a former federal law enforcement officer and firefighter.
http://www.youtube.com/watch?v=48YF1uEuCUA&feature=share

Deconstructing America: Attorneys, Professors, and Judges Speak Out Against CPS-DFCS System
http://www.youtube.com/watch?v=kv42L1XkCBY

This webpage contains information and testimonials from victims of this system.
http://legallykidnapped.blogspot.com

This webpage contains videos, information, links to other organizations, links to new sites, and victim information.
http://amiablyme.wordpress.com/

This video available at YouTube is an Oregon State Senator speaking out against CPS actions that go against the wishes of the legislature of the state, but more importantly it reveals the program that is currently being put in place and will be used as a "national" system.
http://www.youtube.com/watch?v=h4Z-kOLbSg0&feature=share

This is an NBC news report that shows how the "Foster Care System" in coordination with CPS-DFCS agencies are administering psychotropic drugs to very young children. The various researchers who have looked into this issue show that both the CPS-DFCS agencies and the Foster Parents receive increased payments for children under "psychiatric" care. "Drugging also makes the children easier to manage," said a CPS caseworker who resigned from an Oregon office.
http://www.youtube.com/watch?v=nnNyA3v7U80&feature=share

This is a documentary that follows the "institutional abuse of children" in Canada. Be very careful though, the same policies and processes were planned then enacted in the United States and its allies during the 1980's so their problem is the same as ours.
http://www.youtube.com/watch?v=O7horT_3Vy0&feature=related

That documentary explains in detail how the system and the "service providers" work in conspiracy to profit from the destruction of the family.

This link will connect you to the first part of an in-depth report done by a new station in Kentucky. What do events in Kentucky have to do with your state? It is a fact that the agencies in all states are operating in the same way.
http://www.youtube.com/watch?v=jAnjp7OnxNM&feature=related

This is testimony of victims with professionals also included.
http://www.youtube.com/watch?v=wFBv3R1Rsk0&feature=related

The following is a speech that is self-explanatory:

Excerpt, Ted Gunderson Speech Congressional Hearing on Child Protection 3/13/04, San Bernardino California Town Hall Forum w Congressman Joe Baca on Children Protective Services Reform.

"Honorable Lawmakers, Guardians of the United States Constitution and the Federal Treasury, I am a licensed private investigator with more than 54 years' experience which includes more than 27 years as a special agent with the FBI. . .

As a licensed private investigator...Specifically, in regards to Child Protective Services in some areas and some states, I have been told by a reliable source, that a planeload of children from CPS was flown out of Denver, Colorado on Nov 6, 1997 to Paris, France. Later a second plane of children also under the care of CPS was flown from Los Angeles to Europe.

I have also developed information through credible and reliable sources that in the past, children have been taken from Foster Homes, orphanages, and Boys Town Nebraska, and flown by private jets from Sioux City

Iowa to Washington D.C. for sex orgies with politicians.

I have interviewed witnesses who were active in an international child-kidnapping ring, who advised me that, of the thousands of children who disappear every year, many are auctioned off, at various locations throughout the country. This kidnapping ring... One of my sources advised that he has attended six such auctions, with 6 to 36 children being auctioned off.

These locations are identified as fifty miles outside Las Vegas, Nevada, Toronto Canada, Houston Texas, an unidentified location in Michigan and a barn outside Lincoln Nebraska.

This source informed me that the children range in age from 2-21, both boys and girls. They are usually placed on a stage or platform, in their underwear with a number attached to a string around their necks. The perpetrators bid on the children by number. The location outside Las Vegas was at a small airport. Some of the children were auctioned off to foreigners wearing turbans on their heads. The children were placed in private planes from which they took off.

Other children were placed in campers. They were drugged so that if police stopped them the kidnappers could claim their child was sleeping. This same source advised me that when he was ten years old he was used as decoy in public places to attract other children his age to that area, where the adults would grab the kids and flee.

In the early 1990s, following the circulation of "The Franklin Cover-up" for almost a year, the Yorkshire Television of England sent a topnotch investigatory team to produce a documentary. They conducted a national investigation for 10 months, interviewing, filming, and documenting the Franklin story, finding new witnesses, and uncovering new evidence.

Their documentary, "Conspiracy of Silence" was scheduled to be aired nationwide on the Discovery Channel on May 3, 1994 at 10 PM. When certain members of congress learned that Conspiracy of Silence was to be aired on national TV, the cable industry was threatened with restrictive legislation, unknown parties purchased the rights to the documentary and all copies were ordered destroyed.

I visited the LA TIMES LIBRARY, reviewed the TV log for May 3, 1994 noted that Conspiracy of Silence was listed to be aired on that date at 10 PM. I had a contact on the East Coast review the TV log for the Philadelphia Enquirer and Conspiracy of Silence was also listed in their TV section at 10 PM, May 3, 1994. I developed information from a credible source in a major city in Southwest U.S. that there is collusion between judges, attorneys and underworld criminals. Children in the system are adopted for money, the children's names are changed, + each child is re-adopted many times for more money.

The Federal Adoption Bonus is given to these judges, attorneys and underworld criminals it is split among the three groups of child traffickers.
As an outgrowth of my involvement in the Franklin Cover Up Case from Omaha, I learned that a covert CIA operation known as the Finders based in Washington D.C. which was actively involved in kidnapping and international trafficking of children . . ."
Respectfully Presented By THEODORE L. GUNDERSON, FBI Senior Special Agent in Charge (Ret)

Two "youth services providers" can again demonstrate the marriage of the courts, politics, and service providers. The straight Foundation that established and is still connected to at least eight programs operating that use Straight's methods, often in former Straight buildings operated by former Straight staff. They include: Alberta Adolescent Recovery Center (Canada), Pathway Family Center (Michigan, Indiana, Ohio), Growing Together (Florida), Possibilities Unlimited (Kentucky), SAFE (Florida), and Phoenix Institute for Adolescents (Georgia). This organization and its facilities have been connected to Mitt Romney, numerous judges, and many individuals who have considerable influence with the "Family" and "Juvenile" Court systems.

WWASPS is linked with facilities Academy at Ivy Ridge (New York), Carolina Springs Academy (South Carolina), Cross Creek Programs (Utah), Darrington Academy (Georgia), Horizon Academy (Nevada), Majestic Ranch Academy (Utah), MidWest Academy (Iowa), Respect Camp (Mississippi), Royal Gorge Academy (Colorado), Spring Creek Lodge (Montana), and Tranquility Bay (Jamaica).

The irony is that these programs established and touted to be of benefit to youth who are on the boarder of more serious issues, have been found to be abusive in ways that parents do not even want to imagine. U.S. State Department insisted on investigation of the Paradise Cove program in 1998 after receiving "credible allegations of physical abuse," including "beatings, isolation, food and water deprivation, choke-holds, kicking, punching, bondage, spraying with chemical agents, forced medication, verbal abuse and threats of further physical abuse." Paradise Cove closed shortly thereafter.

Yeah, this issue has been going on that long. The marriage financially of the CPS-DFCS, "Family Court", "Juvenile Court", and general judiciary to "privately held agencies that act as service providers to those who are processed through those institutions has been rife with fraud, bribery, kick-backs, and numerous other torte and legal violations. The ease of financial gain and the low risk of any chance of being prosecuted are so disparate that judges and CPS workers have little or no fear of being caught and less fear of being jailed for their crimes," says a national investigative reporter from California. "This is nationwide and has been exported from our country along with the abuse of children and judiciary indiscretion."

Chapter Nine

THE CORRUPT BUSINESS OF CHILD PROTECTIVE SERVICES
BY
Nancy Schaefer
Senator, 50th District, Georgia
Matter of Public Records, State of Georgia

My introduction into child protective service cases was due to a grandmother in an adjoining state who called me with her tragic story. Her two granddaughters had been taken from her daughter who lived in my district. Her daughter was told wrongly that if she wanted to see her children again she should sign a paper and give up her children. Frightened and young, the daughter did. I have since discovered that parents are often threatened into cooperation of permanent separation of their children.

The children were taken to another county and placed in foster care. The foster parents were told wrongly that they could adopt the children. The grandmother then jumped through every hoop known to man in order to get her granddaughters. When the case finally came to court it was made evident by one of the foster parent's children that the foster parents had, at any given time, 18 foster children and that the foster mother had an inappropriate relationship with the caseworker.

In the courtroom, the juvenile judge, acted as though she was shocked and said the two girls would be removed quickly. They were not removed. Finally, after much pressure being applied to the Department of Family and Children Services of Georgia (DFCS), the children were driven to South Georgia to meet their grandmother who gladly drove to meet them. After being with their grandmother two or three days, the judge, quite out of the blue, wrote up a new order to send the girls to their father, who previously had no interest in the case and who lived on the West Coast. The father was in "adult entertainment". His girlfriend worked as an "escort" and his brother, who also worked in the business, had a sexual charge brought against him.

Within a couple of days the father was knocking on the grandmother's door and took the girls kicking and screaming to California.

The father developed an unusual relationship with the former foster parents and soon moved back to the southeast, and the foster parents began driving to the father's residence and picking up the little girls for visits. The oldest child had told her mother and grandmother on two different occasions that the foster father molested her.

To this day after five years, this loving, caring blood relative grandmother does not even have visitation privileges with the children. The little girls are in my opinion permanently traumatized and the young mother of the girls was so traumatized with shock when the girls were first removed from her that she has not recovered.

Throughout this case and through the process of dealing with multiple other mismanaged cases of the Department of Family and Children Services (DFCS), I have worked with other desperate parents and children across the state because they have no rights and no one with whom to turn. I have witnessed ruthless behavior from many caseworkers, social workers, investigators, lawyers, judges, therapists, and others such as those who "pick up" the children. I have been stunned by what I have seen and heard from victims all over the state of Georgia.

In this report, I am focusing on the Georgia Department of Family and Children Services (DFCS). However, I believe Child Protective Services nationwide has become corrupt and that the entire system is broken almost beyond repair. I am convinced parents and families should be warned of the dangers.

The Department of Child Protective Services, known as the Department of Family and Children Service (DFCS) in Georgia and other titles in other states, has become a "protected empire" built on taking children and separating families. This is not to say that there are not those children who do need to be removed from wretched situations and need protection. This report is concerned with the children and parents caught up in "legal kidnapping," ineffective policies, and DFCS who do does not remove a child or children when a child is enduring torment and abuse. (See Exhibit A and Exhibit B)

In one county in my District, I arranged a meeting for thirty-seven families to speak freely and without fear. These poor parents and grandparents spoke of their painful, heart wrenching encounters with DFCS. Their suffering was overwhelming. They wept and cried. Some did not know where their children were and had not seen them in years. I had witnessed the "Gestapo" at work and I witnessed the deceitful conditions under which children were taken in the middle of the night, out of hospitals, off of school buses, and out of homes. In one county a private drug testing business was operating within the DFCS department that required many, many drug tests from parents and individuals for profit. In another county children were not removed when they were enduring the worst possible abuse. Due to being exposed, several employees in a particular DFCS office were fired. However, they have now been rehired either in neighboring counties or in the same county again. According to the calls I am now receiving, the conditions in that county are returning to the same practices that they had before the light was shown on their deeds. Having worked with probably 300 cases statewide, I am convinced there is no responsibility and no accountability in the system.

I have come to the conclusion:

· that poor parents often times are targeted to lose their children because they do not have the where-with-all to hire lawyers and fight the system. Being poor does not mean you are not a good parent or that you do not love your child, or that your child should be removed and placed with strangers;

· that all parents are capable of making mistakes and that making a mistake does not mean your children are always to be removed from the home. Even if the home is not perfect, it is home; and that's where a child is the safest and where he or she wants to be, with family;

· that parenting classes, anger management classes, counseling referrals, therapy classes and on and on are demanded of parents with no compassion by the system even while they are at work and while their children are separated from them. This can take months or even years and it emotionally devastates both children and parents. Parents are victimized by "the system" that makes a profit for holding children longer and "bonuses" for not returning children;

· that caseworkers and social workers are oftentimes guilty of fraud. They withhold evidence. They fabricate evidence and they seek to terminate parental rights. However, when charges are made against them, the charges are ignored;

· that the separation of families is growing as a business because local governments have grown accustomed to having taxpayer dollars to balance their ever-expanding budgets;

· that Child Protective Service and Juvenile Court can always hide behind a confidentiality clause in order to protect their decisions and keep the funds flowing. There should be open records and "court watches"! Look who is being paid! There are state employees, lawyers, court investigators, court personnel, and judges. There are psychologists, and psychiatrists, counselors, caseworkers, therapists, foster parents, adoptive parents, and on and on. All are looking to the children in state custody to provide job security. Parents do not realize that social workers are the glue that holds "the system" together that funds the court, the child's attorney, and the multiple other jobs including DFCS's attorney.

· that The Adoption and the Safe Families Act, set in motion by President Bill Clinton, offered cash "bonuses" to the states for every child they adopted out of foster care. In order to receive the "adoption incentive bonuses" local child protective services need more children. They must have merchandise (children) that sell and you must have plenty of them so the buyer can choose. Some counties are known to give a $4,000 bonus for each child adopted and an additional $2,000 for a "special needs" child. Employees work to keep the federal dollars flowing;

· that there is **double dipping**. The funding continues as long as the child is out of the home. When a child in foster care is placed with a new family then "adoption bonus funds" are available. When a child is placed in a mental health facility and is on 16 drugs per day, like two children of a constituent of mine, more funds are involved;

· that there are no financial resources and no real drive to unite a family and help keep them together;

· that the incentive for social workers to return children to their parents quickly after taking them has disappeared and who in protective services will step up to the plate and say, "This must end!" No one, because they are all in the system together and a system with no leader and no clear policies will always fail the children. Look at the waste in government that is forced upon the tax payer;

· that the "Policy Manuel" is considered "the last word" for DFCS. However, it is too long, too confusing, poorly written and does not take the law into consideration;

· that if the lives of children were improved by removing them from their homes, there might be a greater need for protective services, but today all children are not always safer. Children, of whom I am aware, have been raped and impregnated in foster care and the head of a Foster Parents Association in my District was recently arrested because of child molestation;

· that some parents are even told if they want to see their children or grandchildren, they must divorce their spouse. Many, who are under privileged, feeling they have no option, will divorce and then just continue to live together. This is an anti-family policy, but parents will do anything to get their children home with them.

· fathers, (non-custodial parents) I must add, are oftentimes treated as criminals without access to their own children and have child support payments strangling the very life out of them;

· that the Foster Parents Bill of Rights does not bring out that a foster parent is there only to care for a child until the child can be returned home. Many Foster Parents today use the Foster Parent Bill of Rights to hire a lawyer and seek to adopt the child from the real parents, who are desperately trying to get their child home and out of the system;

· that tax dollars are being used to keep this gigantic system afloat, yet the victims, parents, grandparents, guardians and especially the children, are charged for the system's services.

· that grandparents have called from all over the State of Georgia trying to get custody of their grandchildren. DFCS claims relatives are contacted, but there are cases that prove differently. Grandparents who lose their grandchildren to strangers have lost their own flesh and blood. The children lose their family heritage and grandparents, and parents too, lose all connections to their heirs.

· that The National Center on Child Abuse and Neglect in 1998 reported that six times as many children died in foster care than in the general public and that once removed to official "safety", these children are far more likely to suffer abuse, including sexual molestation than in the general population.

· That according to the California Little Hoover Commission Report in 2003, 30% to 70% of the children in California group homes do not belong there and should not have been removed from their homes.

FINAL REMARKS

On my desk are scores of cases of exhausted families and troubled children. It has been beyond me to turn my back on these suffering, crying, and sometimes beaten down individuals. We are mistreating the most innocent. Child Protective Services have become adult centered to the detriment of children. No longer is judgment based on what the child needs or who the child wants to be with or what is really best for the whole family; it is some adult or bureaucrat who makes the decisions, based often on just hearsay, without ever consulting a family member, or just what is convenient, profitable, or less troublesome for a director of DFCS.

I have witnessed such injustice and harm brought to these families that I am not sure if I even believe reform of the system is possible! The system cannot be trusted. It does not serve the people. It obliterates families and children simply because it has the power to do so. Children deserve better. Families deserve better. It's time to pull back the curtain and set our children and families free.

"Speak up for those who cannot speak for themselves, for the rights of all who are destitute.

Speak up and judge fairly; defend the rights of the poor and the needy." Proverbs 31:8-9

Please continue to read:

Recommendations
Exhibit A
Exhibit B

RECOMMENDATIONS

1. Call for an independent audit of the Department of Family and Children's Services (DFCS) to expose corruption and fraud.

2. Activate immediate change. Every day that passes means more families and children are subject to being held hostage.

3. End the financial incentives that separate families.

4. Grant to parents their rights in writing.

5. Mandate a search for family members to be given the opportunity to adopt their own relatives.

6. Mandate a jury trial where every piece of evidence is presented before removing a child from his or her parents.

7. Require a warrant or a positive emergency circumstance before removing children from their parents. (Judge Arthur G. Christean, Utah Bar Journal, January, 1997 reported that "except in emergency circumstances, including the need for immediate medical care, require warrants upon affidavits of probable cause before entry upon private property is permitted for the forcible removal of children from their parents.")

8. Uphold the laws when someone fabricates or presents false evidence. If a parent alleges fraud, hold a hearing with the right to discovery of all evidence.

Senator Nancy Schaefer
50th District of Georgia

EXHIBIT A
December 5, 2006
Jeremy's Story
(Some names withheld due to future hearings)

As told to Senator Nancy Schaefer by Sandra (XXXX), a foster parent of Jeremy for 2 +½ years.

My husband and I received Jeremy when he was 2 weeks old and we have been the only parents he has really ever known. He lived with us for 27 months. (XXXX) is the grandfather of Jeremy, and he is known for molesting his own children, for molesting Jeremy and has been court ordered not to be around Jeremy. (XXXX) is the mother of Jeremy, who has been diagnosed to be mentally ill, and also is known to have molested Jeremy. (XXXX) and Jeremy's uncle is a registered sex offender and (XXXX) is the biological father, who is a drug addict and alcoholic and who continues to be in and out of jail. Having just described Jeremy's world, all of these adults are not to be any part of Jeremy's life, yet for years DFCS has known that they are. DFCS had to test (XXXX) (the grandfather) and his son (XXXX) (the uncle) and (XXXX) to determine the real father. (XXXX) is the biological father although any of them might have been. In court, it appeared from the case study, that everyone involved knew that this little boy had been molested by family members, even by his own mother, (XXXX). In court, (XXX), the mother of Jeremy, admitted to having had sex with (XXXX) (the grandfather) and (XXXX) (her own brother) that morning. Judge (XXXX) and DFCS gave Jeremy to his grandmother that same day. (XXXX), the grandmother, is over 300 lbs., is unable to drive, and is unable to take care of Jeremy due to physical problems. She also has been in a mental hospital several times due to her behavior. Even though it was ordered by the court that the grandfather (XXXX), the uncle (XXXX) (a convicted sex offender), (XXXX) his mother who molested him and (XXXX) his biological father, a convicted drug addict, were not to have anything to do with the child, they all continue to come and go as they please at (XXXX address), where Jeremy has been sentenced to live for years. This residence has no bathroom and little heat. The front door and the windows are boarded. (See pictures) This home should have been condemned years ago. I have been in this home. No child should ever have to live like this or with such people. Jeremy was taken from us at age 2 +½ years after (XXXX) obtained attorney (XXXX), who was the same attorney who represented him in a large settlement from an auto accident. I am told, that attorney (XXXX), as grandfather's attorney, is known to have repeatedly gotten (XXXX) off of several criminal charges in White County. This is a matter of record and is known by many in White County. I have copies of some records. (XXXX grandfather), through (XXXX attorney's) work, got (XXXX), the grandmother of Jeremy, legal custody of Jeremy. (XXXX grandfather) who cannot read or write also got his daughter (XXXX) and son (XXXX) diagnosed by government agencies as mentally ill. (XXXX grandfather), through legal channels, has taken upon himself all control of the family and is able to take possession of any government funding coming to these people.

It was during this time that Jeremy was to have a six-month transitional period between (XXXX grandmother) and my family as we were to give him up. The court ordered agreement was to have been 4 days at our house and 3 days at (XXXX grandmother). DFCS stopped the visits within 2 weeks. The reason given by DFCS was the child was too traumatized going back and forth. In truth, Jeremy begged us and screamed never to be taken back to (XXXX his grandmother) house, which we have on video. We, as a family, have seen Jeremy in stores time to time with (XXXX grandmother) and the very people he is not to be around. At each meeting Jeremy continues to run to us wherever he sees us and it is clear he is suffering. This child is in a desperate situation and this is why I am writing, and begging you Senator Schaefer, to do something in this child's behalf. Jeremy can clearly describe in detail his sexual molestation by every member of this family and this sexual abuse continues to this day.

When Jeremy was 5 years of age I took him to Dr. (XXXX) of Habersham County who did indeed agree that Jeremy's rectum was black and blue and the physical damage to the child was clearly a case of sexual molestation.

Early in Jeremy's life, when he was in such bad physical condition, we took him to Egleston Children Hospital where at two months of age therapy was to begin three times a week. DFCS decided that the (XXXX grandparent family) should participate in his therapy. However, the therapist complained over and over that the (XXXX grandparent family) would not even wash their hands and would cause Jeremy to cry during these sessions. (XXXX the grandmother), after receiving custody no longer allowed the therapy because it was an inconvenience. The therapist reported that this would be a terrible thing to do to this child. Therapy was stopped and it was detrimental to the health of Jeremy. During (XXXX grandmother) custody, (XXXX uncle) has shot Jeremy with a BB gun and there is a report at (XXXX) County Sheriff's office. There are several amber alerts at Cornelia Wal-Mart, Commerce Wal-Mart, and a 911 report from (XXXX) County Sheriff's Department when Jeremy was lost. (XXXX grandmother), to teach Jeremy a lesson, took thorn bush limbs and beat the bottoms of his feet. Jeremy's feet got infected and his feet had to be lanced by Dr. (XXXX). Then Judy called me to pick him up after about 4 days to take back him to the doctor because of intense pain. I took Jeremy to Dr. (XXXX) in Gainesville. Dr. (XXXX) said surgery was needed immediately and a cast was added. After returning home, (XXXX), his grandfather and (XXXX), his uncle, took him into the hog lot and allowed him to walk in the filth.

Jeremy's feet became so infected for a 2nd time that he was again taken back to Dr. (XXXX) and the hospital. No one in the hospital could believe this child's living conditions. Jeremy is threatened to keep quiet and not say anything to anyone. I have videos, reports, arrest records and almost anything you might need to help Jeremy. Please call my husband, Wendell, or me at any time.

Sandra and (XXXX) husband (XXXX)

EXHIBIT B
Failure of DFCS to remove six desperate children
A brief report regarding six children that Habersham County DFCS director failed to remove as disclosed to Senator Nancy Schaefer by Sheriff Deray Fincher of Habersham County.

Sheriff Deray Fincher, Chief of Police Don Ford and Chief Investigator Lt. Greg Bowen Chief called me to meet with them immediately, which I did on Tuesday, October 16, 2007 Sheriff Fincher, after contacting the Director of Habersham County DFCS several times to remove six children from being horribly abused, finally had to get a court order to remove the children himself with the help of two police officers.

The children, four boys and two girls, were not just being abused; they were being tortured by a monster father.

The six children and a live in girl friend were terrified of this man, the abuser. The children never slept in a bed, but always on the floor. The place where they lived was unfit for human habitation.

The father on one occasion hit one of the boys across his head with a bat and cut the boy's head open. The father then proceeded to hold the boy down and sew up the child's head with a needle and red thread. However, even with beatings and burnings, this is only a fraction of what the father did to these children and to the live-in girlfriend.

Sheriff Fincher has pictures of the abuse and condition of one of the boys and at the writing of this report; he has the father in jail in Habersham County.

It should be noted that when the DFCS director found out that Sheriff Fincher was going to remove the children, she called the father and warned him to flee.

This is not the only time this DFCS director failed to remove a child when she needed to do so. (See Exhibit A)

The egregious acts and abhorrent behavior of officials who are supposed to protect children can no longer be tolerated.

Senator Nancy Schaefer
50th District of Georgia
(originally posted 12/5/07 at FightCPS.Com, a matter of public domain record in the Georgia State Assembly)

INDEX I
Social Work Degree Requirements

The following are the "courses" required to earn a degree in legitimate educational institutions. The first is from the Southeast, 2nd is from Midwest, and the last is from the West Coast.

Example #1.....Compete the following program (major) requirements:
- SWK F103--Introduction to Social Work--3 credits
- SWK F220--Ethics, Values and Social Work Practice--3 credits
- SWK F305O--Social Welfare History--3 credits
- SWK F306--Social Welfare: Policies and Issues--3 credits
- SWK F320W--Rural Social Work--3 credits
- SWK F341--Human Behavior in the Social Environment I--3 credits
- SWK F342--Human Behavior in the Social Environment II--3 credits
- SWK F375W--Research Methods in Social Work--3 credits
- SWK F460--Social Work Practice I--3 credits
- SWK F461--Practicum in Social Work I**--3 or 6 credits
- SWK F463--Social Work Practice II--3 credits
- SWK F464--Practicum in Social Work II**--3 or 6 credits
- SWK F466--Practicum in Social Work III**--3 or 6 credits
- Complete two courses from the following special problems areas:
- HUMS F205--Basic Principles of Group Counseling--3 credits

- HUMS F305--Substance Abuse Counseling--3 credits
- SWK F310--Fetal Alcohol Spectrum Disorder--3 credits
- SWK F330--Seminar in International Social Work--3 credits
- SWK F350W--Women's Issues in Social Welfare and Social Work Practice--3 credits
- SWK F360--Child Abuse and Neglect--3 credits
- SWK F370--Services and Support for an Aging Society--3 credits
- SWK F470--Substance Abuse Theories and Treatment--3 credits
- SWK F484--Seminar in Social Work Practice Areas--3 credits

Example #2....Program Course Requirements and Electives: Total program hours: 28 hours

MAJOR COURSE REQUIREMENTS
SOC 1010 Introduction to Sociology
SPN 1020 Beg Spanish II 5
SW 1010 Introduction to Social Work
SW 2100 Human Behavior in the Social Env
SW 2230 Introduction Group Therapy
SW 2280 Drug/Alcohol Abuse
SW 2650 Social Welfare Policy
SW 2720 Mental Health
SW 2750 Ethics Social Work
SW 2920 1st Yr Social Work Internship
SW 2930 2nd Yr Social Work Internship
Elective 3
ELECTIVES (3 CREDITS)
ETHS 2500 Native American Culture

ETHS 2580 Asian American Culture
ETHS 2660 African American Culture
ETHS 2670 Mexican American Culture
SW 1900 Special Studies 1-2 SW

Example #3
Course No.   Credits
First Term
WR121 English Composition I   4
MTH243 Probability and Statistics   4
PSY201 General Psychology I   4
SOC204 Introduction to Sociology   4
Second Term
WR122 English Composition II   4
SP111 Fundamentals of Public Speaking 4
SOC205 American Society   4
PSY202 General Psychology II   4
CS120 Concepts in Computing I or documented proficiency 0-4
Third Term
ANTH110 Introduction to Cultural Anthropology 4
PSY215 Life Span Human Development 4
— Humanities requirement   4
— Science requirement   3-5
Fourth Term
WR227 Technical Writing   4
BI101 Introduction to Biology I w/lab 4
— Humanities requirement   4
— Health/physical education requirement 3
— Electives to meet 90 credit requirement variable
Fifth Term
SOC213 Multicultural America (recommended) 4
BI102 Introduction to Biology II w/lab 4

— Humanities requirement 4
— Electives to meet 90 credit requirement variable
Sixth term
SOC211 Social Deviance and Social Control 3
— Humanities requirement 4
— Science requirement 4-5
— Electives to meet 90 credit requirement variable
Recommended electives include:
SOC225 Social Problems
SOC230 Gerontology
SOC243 Drugs, Crime and Addiction

INDEX II
Other Forms of Social Work Degrees

Unfortunately, we live in a time of information availability and enormous internet capabilities. However, I only say this because there are many Social Service Workers who have done nothing more than contribute $500-$7500 to an online entity and answer a serious of personal questions. Some have a moderately sized text that must be read, but have no means to discern whether you have read the books they sent you. The book fees are the other profit these illegitimate "educational facilities" make by selling you a diploma. It has been estimated that nearly half of the degrees in CPS agency offices around the United States are degrees of this kind.

The second type of "online course" that is part of that 50% of the degrees inside the CPS offices are those of the kind below. You are required to take 50-60 credits from this online grouping, but some have no testing of any kind. Others have testing, but with auto-correcting answers and no grading system what-so-ever. If you do the work, right-wrong-or-horrendously you still get the "degree" when you complete all the work they set aside for you.

Social Work Concentration in "online program"....
General Education (36 credits)
Additional concentration credits suggested are listed below: (19 credits)
   BIOL 220 Human Anatomy & Physiology (4)
   PHIL 101 Introduction to Philosophy (3)
   PSYC 111 Introduction to Psychology (3)
   SOCI 110 Introduction to Sociology (3)
   SWK 255 Social work in the Modern Society (3)
   SWK 257 Human Behavior in the Social Environment (3)

People with this miniscule amount of education in ANY field are making life altering, life destroying decisions inside CPS offices daily. They are at the least not qualified and at worst, receive training that is both contrary to their assigned and assessed purposes as well as detrimental to the well-being of children and families.

Notes

Made in the USA
Lexington, KY
19 January 2016

AutoCAD LT R2 for Windows
COORDINATE 💻 MODIFERS

COORDINATE INPUT
Default directions:
X-positive: Right
Y-positive: Up
Z-positive: Right-hand rule

x,y	2D cartesian coordinates:	
	From point: **2,3.1**	
x,y,z	3D cartesian coordinates:	
	From point: **2,3,4**	
d<a	2D polar coordinates:	
	From point: **2<45**	
@d<a	Relative coordinates:	
	From point: **@12<34**	
d<a,z	3D cylindrical coordinates:	
	From point: **5<45,5**	
@d<a,z	Relative cylindrical coordinates:	
	From point: **@5<45,5**	
d<a<a	3D spherical coordinates:	
	From point: **5<45<30**	

DISTANCE MODIFIERS

,	Distance separator:	
	From point: **2,3**	
'	Feet (required)	
"	Inches (optional):	
	From point: **2'3,5'4"**	
— and /	Fractional inches:	
	From point: **2'3-3/4,5'4"**	
@	Relative distance (see above)	
*	Force use of WCS coordinates:	
	From line: ***3,4,2**	
@ *	Force relative WCS coordinates:	
	From line: **@*3<45**	

SURVEYOR'S UNITS
E	East	0 degrees
	North	90 degrees
		180 degrees
		⁻⁰ degrees

ANGLE MODIFIERS
Angle defaults:
0-degrees: East (3 o'clock).
Positive direction: Counterclockwise.

'	Minutes
"	Seconds
.	Decimal of a degree or seconds
d	Degrees
r	Radian
g	Grad
<	Angle.
< <	Force 0 degrees=East and decimal degrees
< < <	Force 0=East but use current degree format
+	Counterclockwise
—	Clockwise

OBJECT SNAP MODES
CEN	Center
END	Endpoint
INS	Insertion point
INT	Intersection
MID	Midpoint
NEA	Nearest
NOD	Node (point)
NON	None
PER	Perpendicular
QUA	Quadrant
TAN	Tangent

Modes:
OFF	Turn off osnap
QUI	Quick

POINT FILTERS
.x	pick x, enter y,z
.y	pick y, enter x,z
.z	pick z, enter x,y
.xy	pick xy, enter z
.xz	pick xz, enter y
.yz	pick yz, enter x

Ortho	194
QText	228
Snap	252
Style	256
Toolbox	264
Units	273

V

View Menu

Blipmode	22
Dim: Redraw	100
DView	122
Hide	155
MSpace	180
MView	182
Pan	198
Plan	207
PSpace	224
Redraw	232
Regen	233
Shade	250
View	279
VpLayer	280
VPoint	281
VPorts *or* ViewPorts	282
Zoom	294

W

Windows Related Commands

CopyClip	38
CopyEmbed	39
CopyImage	40
CopyLink	41
Paste Command	200
PasteClip	202
SaveDIB	242
WmfIn	288
WmfOpts	290
WmfOut	291

Quick Start *cont'd*

Designing an Icon Menu 312
Diesel via ModeMacro 315
Enabling Paper Space 225
Exporting Attributes from the Drawing 16
Inserting an AutoCAD LT Object 276
Pasting Text In the Drawing 201
Reloading a Menu File 306
Spell check drawing text 126
Using Isometric Mode 162
Writing a Macro 301
Writing a Script File 246

S

Settings Menu
Aperture 2
AttDisp 12
Base 18
Color *or* Colour 35
DdEModes 49
DdGrips 52
DDim 54
DdLModes 59
DdPtype 68
DdRModes 70
DdSelect 73
DdUnits 79
Dim: Restore 101
Dim: Save 103
Dim: Status 104
Dim: Style 106
Dim: Variables 111
DsViewer 118
Elev 128
Fill 137
Grid 143
Handles 145
Isoplane 161
Layer 164
Limits 166
Linetype 169
[Scale] 175

S

Scripts
- Delay 81
- Resume 235
- RScript 239
- Script 245

T

Text
- AutoCAD SHX Fonts 258
- Change 30
- DdChProp 46
- DdEdit 48
- DdModify 61
- DText 120
- Justification Modes 260
- Paste Command 200
- PostScript PFB Fonts 259
- QText 228
- RevDate 236
- Style 256
- Text 260

Toolbar/Toolbox
- Aerial View toolbar icons 119
- Aliases and Icons 343
- Changing a Toolbar/Toolbox Button 300
- Controlling Layers from the Toolbar 164
- Object Snap Icons 195
- Programming the Toolbar/Toolbox 297
- Toolbox Command 264

Q

Quick Starts
- Advanced Menu Tricks 310
- Changing a Toolbar/Toolbox Button 300
- Converting Slides into Icons 313
- Creating a Custom Linetype 171
- Creating a Simple Menu 305

Modify Menu *cont'd*

    Dim: Override 98
    Dim: TEdit 107
    Dim: TRotate 108
    Dim: Update 110
    Erase 132
    Explode 133
    Extend 134
    HatchEdit 150
    Move 178
    Oops 191
    PEdit 204
    Purge 226
    Rename 234
    RevDate 236
    Rotate 238
    Scale 244
    Stretch 254
    Trim 267

N

New Commands in LT Release ❷

    BHatch 19
    Boundary 27
    DdModify 61
    Divide 114
    Extend 134
    FileOpen 136
    GetEnv 140
    HatchEdit 150
    Measure 176
    RevDate 236
    Trim 267

P

Programming LT

    Aclt.Ini Variables 334
    GetEnv 140
    SetEnv 248
    SetVar 249
    System Variables 321

H

Hatching Commands
BHatch 19
Boundary 27
DdModify 61
Hatch 146
Hatch Pattern Library 147
HatchEdit 150

Help Menu
About 1
Help *or* '? 152

I

Import Into AutoCAD
ASCII text via Clipboard (Paste Command) 200
DWG, AutoCAD file format (Insert) 158
DXF, Autodesk interchange file format (DxfIn) 124
SLD, AutoCAD slide file format (VSlide) 285
WMF, Windows metafile format:
    Clipboard (PasteClip) 202
    File (WmfIn) 288
    Import options (WmfOpts) 290

M

Menu Macro Programming 302

Modify Menu
AttEdit 13
Break 28
Change 30
ChProp 32
DdAttE 44
DdChProp 46
DdEdit 48
DdModify 61
DdRename 69
Dim: Hometext 92
Dim: Newtext 95
Dim: Oblique 96

Export AutoCAD Drawing *cont'd*

```
    Object updating (Update) ................................................. 276
PCX, PC Paintbrush file format (Plot) .................................. 210
PostScript (PsOut) ................................................................ 221
SLD, slide file format (MSlide) .......................................... 179
Text window (LogFileOn) ................................................. 174
TIFF, Adobe tagged image file format (Plot) ................... 210
WMF, Windows metafile format (WmfOut) ..................... 291
Windows Clipboard:
    Bitmap format (CopyImage) ............................................ 40
    OLE embed (CopyEmbed) .............................................. 39
    OLE link (CopyLink) ....................................................... 41
    Picture format (CopyClip) .............................................. 38
```

F

File Menu

```
    AttExt ..................................................................................... 15
    DdAttExt ............................................................................... 45
    DxfIn .................................................................................... 124
    DxfOut ................................................................................. 125
    End ...................................................................................... 131
    FileOpen .............................................................................. 136
    LogFileOff ........................................................................... 173
    LogFileOn ............................................................................ 174
    MSlide .................................................................................. 179
    New ..................................................................................... 184
    Open .................................................................................... 192
    Plot ...................................................................................... 210
    Preferences ......................................................................... 218
    PsOut ................................................................................... 221
    QSave .................................................................................. 227
    Quit ..................................................................................... 229
    Save ..................................................................................... 240
    SaveAs ................................................................................. 241
    SaveDIB ............................................................................... 242
    Script ................................................................................... 245
    Unlock ................................................................................. 275
    Update ................................................................................. 276
    VSlide .................................................................................. 285
    WBlock ................................................................................ 287
    WmfIn .................................................................................. 288
    WmfOpts ............................................................................. 290
    WmfOut ............................................................................... 291
```

DText 120
Ellipse 129
Hatch 146
Insert 158
Line 168
PLine 208
Point 215
Polygon 216
Rectang 230
Solid 253
Text 260
XBind 292
Xref 293
3dPoly 296

E

Edit Menu
CopyClip 38
CopyEmbed 39
CopyImage 40
CopyLink 41
Dim: Undo 109
GraphScr 142
Paste Command 200
PasteClip 202
Redo 231
TextScr 262
U 268
Undo 272

Export AutoCAD Drawing
BMP, Bitmap file format:
    Plot 210
    SaveDIB 242
Data:
    AttExt 15
    DdAttExt 45
DXF, Autodesk drawing interchange file format (DxfOut) 125
GIF, CompuServe graphics interchange file format (Plot) 210
HPGL, Hewlett-Packard file format (Plot) 210
OLE, Object linking and embedding:
    Object embedding (CopyEmbed) 39
    Object linking (CopyLink) 41

C

Construct Menu
- Array 7
- AttDef 10
- Block 23
- BMake 25
- Chamfer 29
- Copy 36
- DdAttDef 42
- Divide 114
- Fillet 139
- Measure 176
- Mirror 177
- Offset 190

D

Diesel Macro Programming 315

Draw Menu
- Arc 3
- BHatch 19
- Boundary 27
- Circle 33
- DdInsert 57
- Dim 82
- Dim: Aligned 85
- Dim: Angular 86
- Dim: Baseline 87
- Dim: Center 88
- Dim: Continue 89
- Dim: Diameter 90
- Dim: Exit 91
- Dim: Horizontal 93
- Dim: Leader 94
- Dim: Ordinate 97
- Dim: Radius 99
- Dim: Rotated 102
- Dim: Vertical 112
- Dim1 84
- DLine 115
- Donut *or* Doughnut 117

Topical Index

A

Aerial View
DsViewer 118
Toolbar icons 119

Assist Menu
Area 6
DdOsnap 67
DdUcs 76
DdUcsP 78
Dist 113
Id 157
List 172
Multiple 181
OSnap 195
Select 247
Time 263
Ucs 269
UcsIcon 271

Attribute Commands
AttDef 10
AttDisp 12
AttEdit 13
AttExt 15
Block 23
BMake 25
DdAttDef 42
DdAttE 44
DdAttExt 45
DdEdit 48
DdInsert 57
DdModify 61
Insert 158

Command	Found In	Notes
X		
XLine	R13	Infinite construction line.
Xplode	R12, R13	More flexible version of the **Explode** command.
XrefClip	R12, R13	Clip a portion of an xref file.
3		
3D	R12, R13	Draw 3D primitive objects.
3dArray	R12, R13	Create a 3D array.
3dFace	R12, R13	Draw a 3D face.
3dMesh	R12, R13	Draw a 3D surface mesh.
3dsIn, 3dsOut	R13	Import, export DWG in 3D Studio format.

Command	Found in	Notes
S		
SaveAsR12	R13	Saves drawing in R12 DWG format.
SaveImg	R12, R13	Saves rendering in raster format.
Scene	R12, R13	Create a scene for rendering.
Section	R12, R13	ACIS solids modelling.
Shape	R12, R13	Place a shape; *use Insert instead.*
Shell	R12, R13	Exit temporarily to operating system; *use [Alt]+[Tab] instead.*
ShowMat	R13	Rendering materials.
Sketch	R12, R13	Freehand drawing.
Slice	R13	ACIS solids modelling.
SOL...	R12	All AME solids modelling with the SOL-prefix.
Spell	R13	Spell check text in drawing.
Sphere	R13	ACIS solids modelling.
Spline, SplinEdit	R13	ACIS solids modelling.
Stats	R12, R13	Rendering statistics.
Status	R12, R13	Status of the drawing.
StlOut	R13	ACIS solids modelling.
Subtract	R13	ACIS solids modelling.
SysWindows	R13	Control multiple windows.
T		
Tablet	R12, R13	Digitizing tablet configuration.
TabSurf	R12, R13	3D surfaces.
TbConfig	R13	Toolbox configuration; *use [Right click] on toolbar icon instead.*
TiffIn	R12, R13	Raster import.
Tolerance	R13	Tolerance dimensioning.
Toolbar	R13	*Use Toolbox command instead.*
TreeStat	R12, R13	Control of drawing hierachy.
U		
Undefine	R12, R13	Undefine a command name.
Union	R13	ACIS solids modelling.
V		
ViewRes	R12, R13	Control roundness of circles.
VlConv	R13	Rendering.
W		
Wedge	R13	ACIS solids modelling.

Command	Found in	Notes
MenuLoad, MenuUnload	R13	Load and unload partial menus.
MInsert	R12, R13	Insert array of blocks; use *Insert* and *Array* instead.
Mirror3D	R12, R13	Mirror about a 3D line.
MlEdit, MLine, MlStyle	R13	Multiline commands.
MText, MtProp	R13	Mutiline text commands.
MvSetup	R12, R13	Drawing setup; use the *New* command instead.

O

Command	Found in	Notes
OleLinks	R13	Control OLE links.

P

Command	Found in	Notes
PasteSpec	R13	Control paste.
PcxIn	R12, R13	Raster import.
PFace	R12, R13	3D faces.
PsDrag, PsFill, PsIn	R12, R13	Control of PostScript images.

R

Command	Found in	Notes
Ray	R13	Semi-infinite construction line.
RConfig	R12, R13	Configure AutoCAD for rendering.
Redefine	R12, R13	Redefine an AutoCAD command.
Redrawall	R12, R13	Redraw all viewports; equivalent to *Redraw* in LT.
Regenall	R12, R13	Regenerate all viewports; equivalent to *Regen* in LT.
RegenAuto	R12, R13	Controls when regnerations occur.
Region	R12, R13	ACIS solids modelling.
Reinit	R12, R13	Reinitializes ports and reloads Acad.Pgp file.
Render, RenderUnload	R12, R13	Rendering.
Replay	R12, R13	Display a raster image.
Revolve	R13	ACIS 3D modelling.
RevSurf	R12, R13	3D surfaces.
RMat	R13	Rendering materials.
Rotate3D	R12, R13	Rotate objects in 3D space.
RPref	R12, R13	Rendering preferences.
RuleSurf	R12, R13	3D surfaces.

Command	Found In	Notes
DxbIn	R12, R13	Import DXB format file.
E		
Edge	R12, R13	Toggle 3D object edge visibility.
EdgeSurf	R12, R13	Draw 3D surface.
Export	R13	Export drawing in non-DWG formats; use *File	Export* instead.
Extrude	R12, R13	3D extrusion command; use *Thickness* system variable instead.
F		
Files	R12, R13	Dialogue box for file management; use Windows *File Manager* instead.
Filmroll	R12	Export drawing in AutoShade FLM format.
Filter	R12, R13	Filter selection set.
G		
GiffIn	R12, R13	Raster import.
Group	R13	Named groups.
H		
HpMPlot	R12, R13	Raster and vector plotting for HPGL/2 plotters.
I		
IgesIn, IgesOut	R12	IGES-format file import and export.
Import	R13	Import drawing in non-DWG formats; use *File	Import* instead.
InsertObj	R13	Place any object from Clipboard into drawing; use *PasteClip* instead.
Interfere	R13	ACIS solids modelling editing command.
Intersect	R13	ACIS solids modelling editing command.
L		
Leader	R13	Use the **Dim:Leader** command instead.
Lengthen	R13	Use the **Extend** and **Trim** command instead.
Light	R12, R13	Places lights for rendering.
Load	R12, R13	Loads shape files; use blocks instead.
M		
MakePreview	R13	Creates preview BMP file of R12 drawings.
MassProp	R13	ACIS solids modelling command.
MatLib	R13	Material library for rendering.
Menu	R12, R13	Load another menu; use *SetEnv MenuName* instead.

Removed Commands

APPENDIX D

As a lower-cost version of AutoCAD, LT has some functionality removed, particularly in the area of 3D drafting, rendering, and solid modelling. The following 193 AutoCAD Release 12 and Release 13 commands are not found in AutoCAD LT Release 2:

Command	Found in	Notes
A		
AcisIn, AcisOut	R13	ACIS solids modelling.
AI_...	R12, R13	All primitive 3D drawing commands prefixed by AI_.
Align	R12, R13	3D drafting.
AmeConvert	R13	ACIS solids modelling.
AppLoad	R12, R13	AutoLISP, ADS, and ARx programming.
Arx	R13	ARx programming.
AscText	R12	ASCII text import; use *Edit	Paste Command* instead.
ASE...	R12, R13	All ASE (AutoCAD SQL Extension) commands prefixed by ASE.
AttReDef	R13	Attribute redefinition.
Audit	R12, R13	Audit drawing file for errors.
B		
BmpOut	R13	Use *SaveDIB* command instead.
Box	R13	ACIS solids modelling.
C		
Cal	R12, R13	Geometry calculator.
Compile	R12, R13	Compile SHP shape and font source code files.
Cone	R13	ACIS solids modelling.
CopyHist	R13	Copy text screen to Clipboard.
CutClip	R13	Cut vectors from drawint to Clipboard.
Cylinder	R13	ACIS solids modelling
D		
DbList	R12, R13	Compete database listing; use *List* command instead.
DdColor	R13	Select color from dialogue box; use *DdRModes* command instead.
DdView	R12, R13	Control named views by dialogue box; use the *View* command instead.
DdVpoint	R12, R13	Select 3D viewpoint by dialogue box; use the *VPoint* command instead.
DIM...	R13	All dimensioning commands prefixed by Dim; use *Dim* command instead.

Command	Alias	GC Alias	Tooltip	Icon #	Icon

Command Modifiers and Point Filters

Command	Alias	GC Alias	Tooltip	Icon #	Icon
Multiple	mu	...	...		
Tracking	tk	...	Tracking	107	
Close	c	...	Close	83	
[Ctrl]+C	\3	...	Cancel	32	^C
@	...	...	@;	61	@
.X	...	...	X	95	.X
.Y	...	...	Y	96	.Y
.Z	...	...	Z	51	.Z

System Variables

Command	Alias	GC Alias	Tooltip
Highlight	...	hi	...
PickBox	pb	...	...
PlineGen	pn	...	...
ShadEdge	sd	...	...
Thickness	th	...	...
Tilemode	tm	...	...

Command	Alias	GC Alias	Tooltip	Icon #	Icon
Object Snap Modes					
Center	cen	...	Center	33	⊙
Endpoint	end	...	Endpoint	46	—
Insertion	ins	...	Insert	54	TEXT
Intersection	int	...	Intersection	55	✕
Midpoint	mid	...	Midpoint	58	—
Nearest	nea	...	...		
Node	nod	...	Node	63	▫
None	non	...	None	64	✳
Perpendicular	per	...	Perpendicular	71	⌐
Quadrant	qua	...	Quadrant	77	⊙
Tangent	tan	...	Tangent	86	◯◯

Command	Alias	GC Alias	Tooltip	Icon #	Icon
Undo	un	oo	Undo	91	↶
Units	ut	nf, un	...		
Unlock	ul	...	...		
V					
View	v	nv, nx	...		
Viewports	vports,vw	vp	...		
VpLayer	vl	...	...		
VSlide	vs	...	...		
W					
WBlock	w	sa	...		
WmfIn	wi	...	...		
WmfOut	wo	...	...		
X					
XBind	xb	...	...		
XRef	xr	...	...		
Z					
Zoom	z	za, zi, zl, zm, zo, zp, zv, zw	Zoom	93	🔍

Command	Alias	GC Alias	Tooltip	Icon #	Icon
Regen	rg	...	...		
Rename	rn	...	...		
Rotate	ro	dr, ro	Rotate	79	

S

Command	Alias	GC Alias	Tooltip	Icon #	Icon
Save	sa	ds	Save	80	
SaveAs	ss	dn	...		
Scale	sc	sz, wz	Scale	81	
Script	sr	...	...		
Select	se	ef, se	...		
Shade	sh	...	...		
Snap	sn	cm, oa, sg	...		
Solid	so	...	...		
Stretch	s	ss, ws	Stretch	85	
Style	st	...	...		

T

Command	Alias	GC Alias	Tooltip	Icon #	Icon
Text	tx	tp	Text	87	
Time	ti	...	...		
Toolbox	tl	...	Toolbox	88	
Trim	tr	it, mt, rm	Trim	89	

U

Command	Alias	GC Alias	Tooltip	Icon #	Icon
Ucs	...	do, mr	Set_UCS	90	
UcsIcon	ui	...	...		

Command	Alias	GC Alias	Tooltip	Icon #	Icon
P					
Pan	p	...	Pan	68	
PasteClip	pc	...	Paste	69	
PEdit	pe	cv	Edit_Polyline	70	
Plan	pv	...	...		
PLine	pl		Polyline	75	
Plot	pp	dp	Print/Plot	72	
Point	pt	po	Point	73	
Polygon	pg	rp	Polygon	74	
Preferences	pf		Preferences	82	
PsOut	pu	...	...		
PSpace	ps	...	...		
Purge	pr	cx, du	...		
Q					
QText	qt	tv	...		
Quit	q,exit,et ...		...		
R					
Rectang	rc	re	Rectangle	30	
Redo	re	uu	Redo	99	
Redraw	r	rd	Redraw	78	

Command	Alias	GC Alias	Tooltip	Icon #	Icon
L					
Layer	la	yc,yd,yh,yn	Layers	56	
Limits	lm	ls	...		
Line	l	li	Line	57	
Linetype	lt	lt	...		
List	ls	...	List	52	
LtScale	lc	lz	...		
M					
Measure	...	...	Measure	105	
Mirror	mi	mi	Mirror	59	
Move	m	mv, wm	Move	60	
MSlide	ml	...	...		
MSpace	ms	...	...		
MView	mv	vs	...		
N					
New	n	dx	New	98	
O					
Offset	of	sl	Offset	65	
Oops	oo	ue	...		
Open	op	...	Open	66	
Ortho	or	or	...		
OSnap	o	...	Object_Snap	67	

Command	Alias	GC Alias	Tooltip	Icon #	Icon
E					
Elev	ev	...	...		
Ellipse	el	ep	Ellipse	45	
Erase	e	el, er, we	Erase	47	
Explode	ep, x	ce, ex	Explode	48	
Extend	ex	mx, xt	Extend	49	
F					
Fill	fl	...	...		
Fillet	f	fi, fr	Fillet	50	
G					
Grid	g	go, gr, gs	...		
H					
Hatch	h	hs, wh	Hatch	97	
HatchEdit	...	...	Edit_Hatch	102	
Help	?	...	Help	109	
Hide	hi	...	...		
I					
Id	...	...	Id	110	
Insert	in	ci, cp	Insert_Block	53	
IsoPlane	is	...	...		

Command	Alias	GC Alias	Tooltip	Icon #	Icon
DdOsnap	os	...	...		
DdRename	dr	...	...		
DdRModes	da	...	Drawing_Aids	100	
DdSelect	sl	...	...		
DdUcs	uc	...	Named_UCS	76	
DdUcsP	up	...	Preset_UCS	108	
DdUnits	du	...	...		
Dim	d	ix,le,lx,rx,ug	Dimension	41	
Dim1	d1	...	...		
Dist	di	me	Distance	84	
Divide	...	...	Divide	106	
DLine	dl	me, db, l2	Double_Line	101	
Donut	do	...	Donut	43	
DsViewer	ds	...	Aerial_View	94	
DText	t, dt	tl	...		
DView	dv	...	...		
DxfIn	dn	...	...		
DxfOut	dx	...	...		

Command	Alias	GC Alias	Tooltip	Icon #	Icon
C					
Chamfer	cf	ca, ch	Chamfer	34	
Change	ch	dg, mp	...		
ChProp	cr	cg, wg	Change_Properties	35	
Circle	c	c2, c3	Circle	36	
Color	co	cs, lc	...		
Copy	cp	co, oc, wc	Copy	37	
CopyClip	cc	...	Copy_Vectors	39	
CopyEmbed	ce	...	...		
CopyImage	ci	...	...		
CopyLink	cl	...	Copy_Link	38	
D					
DdAttDef	dad	...	...		
DdAttE	de	...	...		
DdAttExt	dax	...	...		
DdChProp	dc	...	...		
DdEdit	ed, te	ae, te	Edit_Text	44	
DdEmodes	em	...	Entity_Modes	42	
DdGrips	gr	...	...		
DDim	dm	dd	Dimension_Style	40	
DdInsert	i	...	...		
DdLmodes	ld	...	...		
DdModify	...	tg	Modify_Entity	92	

Command	Alias	GC Alias	Tooltip	Icon #	Icon
...	...	...	O	15	
...	...	...	P	16	
...	...	...	Q	17	
...	...	...	R	18	
...	...	...	S	19	
...	...	...	T	20	
...	...	...	U	21	
...	...	...	V	22	
...	...	...	W	23	
...	...	...	X	24	
...	...	...	Y	25	
...	...	...	Z	26	

A
Command	Alias	GC Alias	Tooltip	Icon #	Icon
About	ab	...	...		
Aperture	ap	to	...		
Arc	a	a2, a3	Arc	27	[icon]
Area	aa	...	Area	111	[icon]
Array	ar	rc	Array	28	[icon]
AttDef	ad	ac, as, at	...		
AttDisp	at	ad	...		
AttEdit	ae	...	...		
AttExt	ax	xa	...		

B
Command	Alias	GC Alias	Tooltip	Icon #	Icon
Base	ba	bp	...		
BHatch			Boundary_Hatch	104	[icon]
BlipMode	bm	pc	...		
Block	b	pc,cc,cn,cw	Block	29	[icon]
Boundary	...	...	Boundary	103	[icon]
Break	br	ob	Break	31	[icon]

Aliases and Icons

APPENDIX C

AutoCAD LT uses a system of aliases and icons to provides command shortcuts. An *alias* is a keyboard abbreviation for a command name, such as 'pr' for the **Preferences** command. An *icon* is a small picture on a toolbar and toolbox button that acts as a shortcut to finding commands on the menu system.

TIPS

■ You get the list of alias names in the Acad.Pgp file, which can be edited with a text editor such as Notepad.

■ The Aclt.Ini initialization file contains the macro and icon definitions of toolbar and toolbox buttons (see Appendix B). The format of a toolbar button, such as **ToolBar1 = \3\3_NEW ^98^**, is as follows:

- **ToolBar1**   Toolbar button # (between 1 and 26).
- **\3\3_NEW**   Macro assigned to button.
- **^98^**   Icon assigned to button.

The format is a toolbox button, such as **ToolBox1=Array ^28^**, is similar:

- **ToolBox1**   Toolbar button # (between 1 and 60).
- **Array**   Macro assigned to button.
- **^28^**   Icon assigned to button.

■ **Command:** AutoCAD LT command name.

■ **Alias:** Command shortcut defined in Aclt.Pgp file.

■ **GC Alias:** Generic CADD commands defined in AcltGcad.Pgp file.

■ **Tooltip:** The name of the icon, tooltip phrase, and toolbar alias.

■ **Icon #:** The icon number used in toolbar/toolbox macros.

■ **Icon:** The icon picture.

Command	Alias	GC Alias	Tooltip	Icon #	Icon
...	...	...	A	1	
...	...	...	B	2	
...	...	...	C	3	
...	...	...	D	4	
...	...	...	E	5	
...	...	...	F	6	
...	...	...	G	7	
...	...	...	H	8	
...	...	...	I	9	
...	...	...	J	10	
...	...	...	K	11	
...	...	...	L	12	
...	...	...	M	13	
...	...	...	N	14	

Variable	Default	Meaning

Win.Ini File

AutoCAD LT-related items in the initialization file for Windows, found in the \Windows subdirectory.

[Extensions]
The association between DWG files and AutoCAD LT:
DWG=C:\Acltwin\Aclt.Exe ^.dwg

[Embedding]
Information used by OLE for embedding DWG file from AutoCAD LT:
AutoCAD-LT=AutoCAD LT Drawing, AutoCAD LT Drawing, C:\Acltwin\Aclt.Exe,picture

Variable	Default	Meaning

AcltDrv.Ini File

Initialization file for AutoCAD LT display driver and Aerial View, found in the \Windows subdirectory.

[INIT]

Driver   GDI   Display driver used by AutoCAD LT:
                                       GDI   Graphics Display Interface.
                                       BpAcLt.Dll   Autodesk display driver.

Driver_Name   GDI   Toggle GDI bypass:
                                        GDI   Use GDI (default).
                                        GDI bypass   Bypass GDI.

[BEV]

BEV is short for bird's-eye view, an alternate name for the Aerial View window.

BarSize   20   Size of icons on Aerial View toolbar:
                                        20   Small icons (default).
                                        30   Large icons.

height   160   Location of Aerial View window, in pixels.
left   278
top   139
width   472

Toolbar   1   Toggle display of toolbar:
                                        0   No toolbar.
                                        1   Toolbar (default).

Bev_Auto_Update   1   Automatically update Aerial View:
                                        0   Don't update automatically.
                                        1   Update automatically.

Variable	Default	Meaning
	ToolBox23	_NEAREST ^62^
	ToolBox24	_NONE ^64^
	ToolBox25	\3\3_COPY ^37^
	ToolBox26	\3\3_ERASE ^47^
	ToolBox27	\3\3_MOVE ^60^
	ToolBox28	\3\3_ROTATE ^79^
	ToolBox29	\3\3_SCALE ^81^
	ToolBox30	\3\3_STRETCH_C ^85^
	ToolBox31	\3\3_BREAK ^31^
	ToolBox32	\3\3_EXTEND ^49^
	ToolBox33	\3\3_TRIM ^89^
	ToolBox34	\3\3_DDCHPROP ^35^
	ToolBox35	\3\3_EXPLODE ^48^
	ToolBox36	\3\3_DDEDIT ^44^
	ToolBox37	'_REDRAW ^78^
	ToolBox38	'_PAN ^68^
	ToolBox39	\3\3_DDRMODES ^100^
	ToolBox40	'@ ^61^
	ToolBox41	.X ^95^
	ToolBox42	.Y ^96^
	ToolBox43	_TRACKING ^107^
	ToolBox44	_CLOSE ^83^
	ToolBox45	'_ID ^110^
	ToolBox46	'_DIST ^84^
	ToolBox47	\3\3_AREA ^111^
	ToolBox48	\3\3_LIST ^52^
	Toolbox49	
	...	
	Toolbox60	

[WMF]

NoFill	0	Toggle import of filled areas:
		  0 No fill, display outline (default).
		  1 Fill areas.

Preview	1	Display preview image in **Open WMF** dialogue box:
		  0 No preview image.
		  1 Display preview image (default).

WideLines	0	Toggle import of wide lines:
		  0 Import wide lines as zero-width lines (default).
		  1 Import wide lines.

Variable	Default	Meaning

[AutoCAD LT ToolBox]

TBoxLocation	680 101	Screen location of Toolbox, in pixels.
TBoxLockedWidth	4	Width of Toolbox when docked, in buttons: 1 Minimum. 4 Default. 60 Maximum.
TBoxOption	3	State of Toolbox: 0 Docked at the right side of LT window. 1 Floating (default). 2 Docked at the left side of the LT window. 3 Turned off.
TBoxStart	1	Toolbox display when LT starts: 0 Off. 1 On (default).
TBoxWidth	12	Width of Toolbox when floating, in buttons: 1 Minimum. 12 Default. 60 Maximum.
ToolBox*nn*		Macro and icon for toolbox button *n*: ^*nn*^ Icon number (see Appendix C)

The first 48 buttons have these defaults:

ToolBox1	\3\3_LINE ^57^
ToolBox2	\3\3_ARC ^27^
ToolBox3	Circle ^36^
ToolBox4	\3\3_DTEXT ^87^
ToolBox5	\3\3_PLINE ^75^
ToolBox6	\3\3_POINT ^73^
ToolBox7	\3\3_BHATCH ^104^
ToolBox8	\3\3_DONUT ^43^
ToolBox9	\3\3_ELLIPSE ^45^
ToolBox10	\3\3_POLYGON ^74^
ToolBox11	\3\3_RECTANG ^30^
ToolBox12	\3\3_DLINE ^101^
ToolBox13	'_DDOSNAP ^67^
ToolBox14	_ENDP ^46^
ToolBox15	_INT ^55^
ToolBox16	_MID ^58^
ToolBox17	_PER ^71^
ToolBox18	_CEN ^33^
ToolBox19	_TAN ^86^
ToolBox20	_QUAD ^77^
ToolBox21	_INS ^54^
ToolBox22	_NODE ^63^

Variable	Default	Meaning

[AutoCAD LT Text Window]

Background	255 255 255	LT text window background color (default = white):
		255 Red (range: 0 to 255).
		255 Green (range: 0 to 255).
		255 Blue (range: 0 to 255).
Font	Courier 10 400 0	Font used by LT text window:
		"Courier" Name of font.
		10 Size of font.
		700 Weight of font:
		400 Regular.
		700 Boldface.
		0 Angle of font:
		0 Regular.
		1 Italic.
Foreground	0 0 0	LT text window text color (default = black):
		255 Red (range: 0 to 255).
		255 Green (range: 0 to 255).
		255 Blue (range: 0 to 255).
Visible	0	Visibility of LT text window:
		0 Off (default).
		1 On.
WindowPosition	Left 13, Top 210, Right 653, Bottom 414	
		Position of LT text window when, in pixels.

Variable	Default	Meaning

[AutoCAD LT Graphics Window]

Font	Arial 10 700 0	Font used by AutoCAD LT window's GUI:
		"Arial" Name of font
		10 Size of font
		700 Weight of font:
		    400 Regular
		    700 Boldface
		0 Angle of font:
		    0 Regular
		    1 Italic

GraphicsBackground	255 255 255	LT window drawing area background color in RGB format (default = white): 255 Red (range: 0 to 255) 255 Green (range: 0 to 255) 255 Blue (range: 0 to 255)
TextBackground	255 255 255	'Command:' prompt area background color in RGB format (default = white): 255 Red (range: 0 to 255) 255 Green (range: 0 to 255) 255 Blue (range: 0 to 255)
TextForeground	0 0 0	'Command:' prompt area text color in RGB format (default = black): 255 Red (range: 0 to 255) 255 Green (range: 0 to 255) 255 Blue (range: 0 to 255)
ToolBar	1	Toggle visibility of toolbar: 0 Off 1 On (default)
WindowPosition	Left 0, Top 0, Right 640, Bottom 480	Position of LT window when **WindowState** = 5, in pixels.
WindowState	5	State of LT window: 3 Maximized 5 Resizeable (default) 6 Minimized (displayed as an icon)
XhairPickboxEtc	0 0 0	Color of crosshair, pickbox, and aperture components of the cursor in RGB format (default = black): 255 Red (range: 0 to 255) 255 Green (range: 0 to 255) 255 Blue (range: 0 to 255)

Variable	Default	Meaning
TempFilesDefault	1	Location to store temporary files: 0 Use subdir specified by **TempFilesDir**. 1 Use drawing subdirectory (default).
TempFilesDir		Name of subdir for temporary files (default = blank)
ToolBar		Macro and icon for toolbar button *n*: ^*nn*^ Icon number (see Appendix C). *The first 48 buttons have these defaults:* ToolBar1 \3\3_NEW ^98^ ToolBar2 \3\3_OPEN ^66^ ToolBar3 \3\3_SAVE ^80^ ToolBar4 \3\3_PLOT ^72^ ToolBar5 _U ^91^ ToolBar6 _REDO ^99^ ToolBar7 '_ZOOM ^93^ ToolBar8 '_HELP ^109^ ToolBar9 '_DDLMODES ^56^ ToolBar10 \3\3_TOOLBOX ^88^ ToolBar11 _DSVIEWER ^94^ ToolBar12 \3\3 ^32^ ToolBar13 ... ToolBar26
ToolBarSize	16	Toolbar and toolbox icon size, in pixels: 4 Minimum. 16 Default. 24 Largest recommended size. 32 Maximum.
UsePreviousPrinter	No	Determines the plotter used: No Use Windows system printer. Yes Use selected printer.
UserName	"Ralph Grabowski, XYZ Publishing, Ltd." Name of registered user; used by **RevDate** command.	

Variable	Default	Meaning

Dither	0	Dither colors for monochrome plotted output:
		  0 Off.
		  1 On.

Drawing1 ... Drawing4		Subdir path and filename of four most-recently loaded DWG files.

FileLocking	1	Toggle file locking:
		  0 Off.
		  1 On (default).

ISOHatch	acltiso.pat	ISO hatch pattern filename.
ISOLinetype	acltiso.lin	ISO linetype filename.
ISOPrototype	acltiso.dwg	ISO prototype filename.

Measure	1	Format of measurement:
		  1 ANSI (imperial).
		  0 ISO (metric).

MenuName	aclt	Menu filename.

MonoVectors	0	Display drawing in monochrome (black):
		  0 Off (default).
		  1 On.

SetupMode	1	Type of drawing setup during **New** command:
		  0 No setup.
		  1 Quick setup (default).
		  2 Custom setup.

ShowStartup	0	Toggle dialogue box display during **New** command:
		  0 **Create New Drawing** dialogue box.
		  1 **Open Drawing** dialogue box.

Support	c:\acltwin	Support subdirectory name.

TBlock	ISO A0 (mm)	Title block filename and description:
		  TBlock1 ansi_a.dwg, ANSI A (in).
		  TBlock2 ansi_b.dwg, ANSI B (in).
		  TBlock3 ansi_c.dwg, ANSI C (in).
		  TBlock4 ansi_d.dwg, ANSI D (in).
		  TBlock5 ansi_e.dwg, ANSI E (in).
		  TBlock6 ansi_v.dwg, ANSI V (in).
		  TBlock7 archeng.dwg, Arch/Eng (in).
		  TBlock8 gs24x36.dwg, Gneric D (in).
		  TBlock9 iso_a0.dwg, ISO A0 (mm).
		  TBlock10 iso_a1.dwg, ISO A1 (mm).
		  TBlock11 iso_a2.dwg, ISO A2 (mm).
		  TBlock12 iso_a3.dwg, ISO S3 (mm).
		  TBlock13 iso_a4.dwg, ISO A4 (mm) .

TBlockRevDate	yes	Run **RevDate** command during **New** command:
		  Yes (yes).
		  No.

Aclt.Ini Variables

APPENDIX B

AutoCAD LT stores information about the current state of itself, the drawing, and the operating system in *system variables*. In addition to the system variables listed in Appendix A, LT stores Windows-specific variables in the Aclt.Ini, AcltDrv.Ini, and Windows.Ini initialization files, as listed in this appendix.

TIPS

■ You get the value of a variable in the **[AutoCAD LT General]** section at the 'Command:' prompt with the **GetEnv** command:
    Command: **getenv**
    Variable name: **monovectors**
    0

■ The **SetEnv** command lets you change the value of variables in — and add variables to — the **[AutoCAD LT General]** section:
    Command: **setenv**
    Variable name: **monovectors**
    New value for MONOVECTORS <0>: **1**

■ Variables in the INI files can be edited directly with a text editor, such as Notepad.

■ Not all variables listed below show up in the initialization files; AutoCAD LT adds variables as required.

Variable	Default	Meaning
[AutoCAD LT General]		
ANSIHatch	aclt.pat	ANSI hatch pattern filename.
ANSILinetype	aclt.lin	ANSI linetype filename.
ANSIPrototype	aclt.dwg	ANSI prototype filename.

AutoSave	0	Toggles automatic back up:
		  0 On.
		  1 Off.

AutoSaveFile	auto.sv$	Subdir path and filename of automatic backup file.

AutoSaveInterval	15	Time between automatic backups, in minutes:
		  1 Minimum.
		  15 Default.
		  600 Maximum.

BeepOnError	0	Toggle beep sound when error occurs:
		  0 Off (default).
		  1 On.

CustomColors	0	User interface colors have been customized in sections **[AutoCAD LT Graphics Window]** and **[AutoCAD LT Text Window]** sections:
		  0 No (default).
		  1 Off.

Variable	Default	Ro	Loc	Meaning
W				
WORLDUCS	1	R/o	...	Matching of WCS with UCS: 0 Current UCS is not WCS. 1 UCS is WCS.
WORLDVIEW	1	...	Dwg	Display during **DView** and **VPoint** commands: 0 Display UCS. 1 Display WCS.
X				
XREFCTL	0	...	Cfg	Creation of XLG xref log files: 0 File no written 1 XLG file written

Variable	Default	Ro	Loc	Meaning

U

Variable	Default	Ro	Loc	Meaning
UCSFOLLOW	0	...	Dwg	New UCS views: 0 No change. 1 Automatic display of plan view.
🖼 UCSICON	1	...	Dwg	Display of UCS icon: 0 Off. 1 On. 2 Display at UCS origin, if possible.
UCSNAME	""	R/o	Dwg	Name of current UCS view: "" Current UCS is unnamed.
UCSORG	0.0000,0.0000,0.0000	R/o	Dwg	Origin of current UCS relative to WCS.
UCSXDIR	1.0000,0.0000,0.0000	R/o	Dwg	X-dir of current UCS relative to WCS.
UCSYDIR	0.0000,1.0000,0.0000	R/o	Dwg	Y-dir of current UCS relative to WCS.
UNDOCTL	5	R/o	...	State of undo: 0 Undo disabled. 1 Undo enabled. 2 Undo limited to one command. 4 Auto-group mode. 8 Group currently active.
UNDOMARKS	0	R/o	...	Current number of undo marks.
UNITMODE	0	...	Dwg	Units display: 0 As set by **Units** command. 1 As entered by user.
USERI1–I5	0	...	...	Five user-definable integer variables.
USERR1–R5	0.0000	...	...	Five user-definable real variables.

V

Variable	Default	Ro	Loc	Meaning
VIEWCTR	6.2433,4.5000,0.0000	R/o	Dwg	X,y-coordinate of center of current view.
VIEWDIR	0.0000,0.0000,1.0000	R/o	Dwg	Current view direction relative to UCS.
VIEWMODE	0	R/o	Dwg	Current view mode: 0 Normal view. 1 Perspective mode on. 2 Front clipping on. 4 Back clipping on. 8 UCS-follow on. 16 Front clip not at eye.
VIEWSIZE	9.0000	R/o	Dwg	Height of current view.
VIEWTWIST	0	R/o	Dwg	Twist angle of current view.
VISRETAIN	0	...	Dwg	Determination of xref drawing's layers: 0 Current drawing. 1 Xref drawing.
VSMAX	47.2227,27.00,0.00	R/o	Dwg	Upper right corner of virtual screen.
VSMIN	-31.8151,-18.00,0.00	R/o	Dwg	Lower left corner of virtual screen.

Variable	Default	Ro	Loc	Meaning
SPLFRAME	0	...	Dwg	Polyline and mesh display: 0 Polyline control frame not displayed; Display polygon fit mesh; 3D faces invisible edges not displayed. 1 Polyline control frame displayed; display polygon defining mesh; 3D faces invisible edges displayed.
SPLINESEGS	8	...	Dwg	Number of line segments that define a splined polyline.
SPLINETYPE	6	...	Dwg	Spline curve type: 5 Quadratic bezier spline. 6 Cubic bezier spline.
SYSCODEPAGE	"iso8859-1"	R/o	Dwg	System code page.

T

Variable	Default	Ro	Loc	Meaning
TARGET	0.0000,0.0000,0.0000	R/o	Dwg	Target in current viewport.
TDCREATE	2448860.54014699	R/o	Dwg	Time and date drawing created.
TDINDWG	0.00040625	R/o	Dwg	Duration drawing loaded.
TDUPDATE	2448860.54014699	R/o	Dwg	Time and date of last update.
TDUSRTIMER	0.00040694	R/o	Dwg	Time elapsed by user-timer.
TEXTEVAL	0	...	...	*Interpretation of text input:* *0 Literal text.* *1 Read (and ! as AutoLISP code.*
TEXTSIZE	0.2000	...	Dwg	Current height of text.
TEXTSTYLE	"STANDARD"	...	Dwg	Current name of text style.
THICKNESS	0.0000	...	Dwg	Current entity thickness.
TILEMODE	1	...	Dwg	Viewport mode: 0 Display tiled viewports. 1 Display overlapping viewports.
TOOLTIPS	1	...	Cfg	Display tooltips: 0 Off. 1 On.
TRACEWID	*0.0500*	...	*Dwg*	*Current width of traces.*
TREEDEPTH	*3020*	...	*Dwg*	*Maximum branch depth in xxyy format:* *xx Model-space nodes.* *yy Paper-space nodes* *+n 3D drawing.* *-n 2D drawing.*
TREEMAX	*10000000*	...	*Cfg*	*Limits memory consumption during drawing regeneration.*

Variable	Default	Ro	Loc	Meaning
S				
SAVEFILE	"AUTO.SV$"	R/o	Cfg	Automatic save filename.
SAVENAME	""	R/o	...	Drawing save-as filename.
SAVETIME	0	...	Cfg	Automatic save interval, in minutes:
				0 Disable auto save.
				120 Default.
SCREENSIZE	*575.0000,423.0000*	*R/o*	...	*Current viewport size, in pixels.*
SHADEDGE	3	...	Dwg	Shade style:
				0 Shade faces (256-color shading).
				1 Shade faces; edges background color.
				2 Simulate hidden-line removal.
				3 16-color shading (default).
SHADEDIF	70	...	Dwg	Percent of diffuse to ambient light:
				0 Minimum.
				70 Default.
				100 Maximum.
SNAPANG	0	...	Dwg	Current rotation angle for snap & grid.
SNAPBASE	0.0000,0.0000	...	Dwg	Current origin for snap and grid.
SNAPISOPAIR	0	...	Dwg	Current isometric drawing plane:
				0 Left isoplane.
				1 Top isoplane.
				2 Right isoplane.
SNAPMODE	0	...	Dwg	Snap mode:
				0 Off.
				1 On.
SNAPSTYL	0	...	Dwg	Snap style:
				0 Normal.
				1 Isometric.
SNAPUNIT	1.0000,1.0000	...	Dwg	X,y-spacing for snap.
SORTENTS	*96*	...	*Cfg*	*Entity display sort order:*
				0 Off.
				1 Object selection.
				2 Object snap.
				4 Redraw.
				8 Slide generation.
				16 Regeneration.
				32 Plots.
				64 PostScript output.

Variable	Default	Ro	Loc	Meaning
PERIMETER	0.0000	R/o	...	Perimeter calculated by **Area** command.
PHANDLE	*0*	...	...	*2,803,348,672 Maximum.*
PICKADD	1	...	Cfg	Effect of **[Shift]** key on selection set:
				0 Adds to selection set.
				1 Removes from selection set.
PICKAUTO	1	...	Cfg	Selection set mode:
				0 Single pick mode.
				1 Automatic windowing & crossing.
PICKBOX	3	...	Cfg	Object selection pickbox size, in pixels.
PICKDRAG	0	...	Cfg	Selection window mode:
				0 Pick two corners.
				1 Pick 1 corner; drag to 2nd corner.
PICKFIRST	1	...	Cfg	Command-selection mode:
				0 Enter command first.
				1 Select objects first.
PLATFORM	"Microsoft Windows"	R/o	Aclt	AutoCAD platform name.
PLINEGEN	0	...	Dwg	Polyline linetype generation:
				0 From vertex to vertex.
				1 From end to end.
PLINEWID	0.0000	...	Dwg	Current polyline width.
PLOTID	""	...	Cfg	Current plotter.
PLOTTER	1	...	Cfg	Current plotter configuration number:
				0 No plotter configured.
				29 Maximum configurations.
POLYSIDES	4	...	...	Current number of polygon sides:
				3 Minimum sides.
				4 Default.
				1024 Maximum sides.
PSLTSCALE	0	...	Dwg	Paper space linetype scaling:
				0 Use model space scale factor.
				1 Use viewport scale factor.
PSPROLOG	""	...	Cfg	PostScript prologue filename.
PSQUALITY	75	...	Dwg	PostScript display resolution in pixels:
				$-n$ Display as outlines; no fill.
				0 No display.
				$+n$ Display filled.

Q

Variable	Default	Ro	Loc	Meaning
QAFLAGS	*0*	...	...	*Quality assurance flags.*
QTEXTMODE	0	...	Dwg	Quick text mode:
				0 Off.
				1 On.

Variable	Default	Ro	Loc	Meaning
MENUECHO	0	...		Menu and prompt echoing: 0 All prompts displayed. 1 Suppress menu echoing. 2 Suppress system prompts. 4 Disable ^P toggle. 8 Display all input-output strings.
MIRRTEXT	1	...	Dwg	Text handling during Mirror command: 0 Mirror text. 1 Retain text orientation.
MODEMACRO	""	...	...	Invoke Diesel programming language.

O

Variable	Default	Ro	Loc	Meaning
OFFSETDIST	-1.0000	...	...	Current offset distance: -1 Through mode, if negative.
ORTHOMODE	0	...	Dwg	Orthographic mode: 0 Off. 1 On.
OSMODE	0	...	Dwg	Current object snap mode: 0 NONe. 1 ENDpoint. 2 MIDpoint. 4 CENter. 8 NODe. 16 QUAdrant. 32 INTersection. 64 INSertion. 128 PERpendicular. 256 TANgent. 512 NEARest. 1024 QUIck.

P

Variable	Default	Ro	Loc	Meaning
PDMODE	0	...	Dwg	Point display mode: 0 Dot. 1 No display. 2 +-symbol. 3 x-symbol. 4 Short line. 32 Circle. 64 Square.
PDSIZE	0.0000	...	Dwg	Point display size, in pixels: -1 Absolute size. 0 5% of drawing area height. +1 Percentage of viewport size.

Variable	Default	Ro	Loc	Meaning
HPNAME	"ANSI31"	...	...	Current hatch pattern name: "" No default. . Set no default.
HPSCALE	1.0000	...	...	Current hatch pattern scale factor.
HPSPACE	1.0000	...	...	Spacing of user-defined hatching.

I

Variable	Default	Ro	Loc	Meaning
INSBASE	0.0000,0.0000,0.0000	...	Dwg	Insertion base point relative to current UCS.
INSNAME	""	...	...	Current block name: . Set to no default. "" No default.

L

Variable	Default	Ro	Loc	Meaning
LASTANGLE	0	R/o	...	Ending angle of last-drawn arc.
LASTPOINT	0.0000,0.0000,0.0000	...	Dwg	Last-entered point.
LENSLENGTH	50.0000	R/o	Dwg	Perspective view lens length, in mm.
LIMCHECK	0	...	Dwg	Drawing limits checking: 0 Disabled. 1 Enabled.
LIMMAX	12.0000,9.0000	...	Dwg	Upper right drawing limits.
LIMMIN	0.0000,0.0000	...	Dwg	Lower left drawing limits.
🖬 LTSCALE	1.0000	...	Dwg	Current linetype scale factor.
LUNITS	2	...	...	Linear units mode: 1 Scientific. 2 Decimal. 3 Engineering. 4 Architectural. 5 Fractional.
LUPREC	4	...	Dwg	Decimal places of linear units.

M

Variable	Default	Ro	Loc	Meaning
MACROTRACE	0	...	...	*Diesel debug mode:* *0 Off.* *1 On.*
MAXACTVP	16	...	...	Maximum viewports to regenerate: 0 Minimum. 16 Default. 32767 Maximum.
MAXSORT	200	...	Cfg	*Maximum filenames to sort alphabetically:* *0 Minimum.* *16 Default.* *32767 Maximum.*
MAXOBJMEM	2,147,483,647	...	...	Max number of objects in memory.

Variable	Default	Ro	Loc	Meaning
F				
FILEDIA	1	...	Cfg	User interface: 0 Command-line prompts. 1 Dialogue boxes (when available).
FILLETRAD	0.5000	...	Dwg	Current fillet radius.
FILLMODE	1	...	Dwg	Fill of solid objects: 0 Off. 1 On.
FLATLAND	*0*	*R/o*	*...*	*Obsolete system variable.*
FRONTZ	0.0000	R/o	Dwg	Front clipping plane offset.
G				
GRIDMODE	0	...	Dwg	Display of grid: 0 Off. 1 On.
GRIDUNIT	0.0000,0.0000	...	Dwg	X,y-spacing of grid.
GRIPBLOCK	0	...	...	Display of grips in blocks: 0 At insertion point. 1 At all entities within block.
GRIPCOLOR	5	...	Cfg	Color of unselected grips: 1 Minimum color number. 5 Default color (blue). 255 Maximum color number.
GRIPHOT	1	...	Cfg	Color of selected grips: 1 Default (red). 255 Maximum color number.
GRIPS	1	...	Cfg	Display of grips: 0 Off. 1 On.
GRIPSIZE	3	...	Cfg	Size of grip box, in pixels: 1 Minimum size. 3 Default size. 255 Maximum size.
H				
🖫 HANDLES	1	R/o	...	Entity handles are being used.
HIGHLIGHT	1	...	...	Object selection highlighting: 0 Disabled. 1 Enabled.
HPANG	0	...	...	Current hatch pattern angle.
HPDOUBLE	0	...	...	Double hatching: 0 Disabled. 1 Enabled.

Variable	Default	Ro	Loc	Meaning
DIMZIN	0		Dwg	Suppression of zero in feet-inches units:
				0 Suppress 0 feet and 0 inches.
				1 Include 0 feet and 0 inches.
				2 Include 0 feet; suppress 0 inches.
				3 Suppress 0 feet; include 0 inches.
DISTANCE	0.0000	R/o	...	Distance measured by **Dist** command.
DONUTID	0.5000	...	...	Inside radius of donut.
DONUTOD	1.0000	...	...	Outside radius of donut.
DWGCODEPAGE	"iso8859-1"		Dwg	Drawing code page.
DWGNAME	"UNNAMED"	R/o	...	Current drawing filename.
DWGPREFIX	"d:\ACLT\"	R/o	...	Drawing's drive and subdirectory.
DWGTITLED	0	R/o	...	Drawing has filename:
				0 "Untitled.Dwg".
				1 User-assigned name.
DWGWRITE	1	...	...	Drawing read-write status:
				0 Read-only.
				1 Read-write.

E

Variable	Default	Ro	Loc	Meaning
ELEVATION	0.0000	...	Dwg	Current elevation relative to current UCS.
ENTMODS	*193*	*R/o*	*...*	*Numer of modified entities.*
ERRNO	*0*	*...*	*...*	*Error number from program.*
EXEDIR	"d:\ACLT\"	...	...	Subdir location of **Aclt.Exe** program.
EXPERT	0	...	...	Controls prompts:
				0 Normal prompts.
				1 Supress these messages: "About to regen, proceed?" "Really want to turn the current layer off?"
				2 Also suppress: "Block already defined.Redefine it?" "A block with this name already exists. Overwrite it?"
				3 Also suppress messages related to the **Linetype** command.
				4 Also suppress messages related to the **UCS Save** and **VPorts Save** commands.
				5 Also suppress messages related to the **DimStyle Save** and **DimOverride** commands.
EXTMAX	-1.0E+20,-1.0E+20,-1.0E+20	R/o	Dwg	Upper right coord of drawing extents.
EXTMIN	1.0E+20,1.0E+20,1.0E+20	R/o	Dwg	Lower left coordinate of drawing extents.

Variable	Default	Ro	Loc	Meaning
DIMBLK	""	R/o	Dwg	Arrow block name.
DIMBLK1	""	R/o	Dwg	First arrow block name.
DIMBLK2	""	R/o	Dwg	Second arrow block name.
DIMCEN	0.0900	...	Dwg	Center mark size.
DIMCLRD	0	...	Dwg	Dimension line color.
DIMCLRE	0	...	Dwg	Extension line & leader color.
DIMCLRT	0	...	Dwg	Dimension text color.
DIMDEC	4	...	Dwg	Primary tolerance decimal places.
DIMDLE	0.0000	...	Dwg	Dimension line extension.
DIMDLI	0.3800	...	Dwg	Dimension line continuation increment.
DIMEXE	0.1800	...	Dwg	Extension above dimension line.
DIMEXO	0.0625	...	Dwg	Extension line origin offset.
DIMGAP	0.0900	...	Dwg	Gap from dimension line to text.
DIMLFAC	1.0000	...	Dwg	Linear unit scale factor.
DIMLIM	0	...	Dwg	Generate dimension limits.
DIMPOST	""	...	Dwg	Default suffix for dimension text.
DIMRND	0.0000	...	Dwq	Rounding value.
DIMSAH	0	...	Dwg	Separate arrow blocks.
DIMSCALE	1.0000	...	Dwg	Overall scale factor.
DIMSE1	0	...	Dwg	Suppress the first extension line.
DIMSE2	0	...	Dwg	Suppress the second extension line.
DIMSHO	1	...	Dwg	Update dimensions while dragging.
DIMSOXD	0	...	Dwg	Suppress outside extension dimension.
DIMSTYLE	"*UNNAMED"	R/o	Dwg	Current dimension style.
DIMTAD	0	...	Dwg	Place text above the dimension line.
DIMTFAC	1.0000	...	Dwg	Tolerance text height scaling factor.
DIMTIH	1	...	Dwg	Text inside extensions is horizontal.
DIMTIX	0	...	Dwg	Place text inside extensions.
DIMTM	0.0000	...	Dwg	Minus tolerance.
DIMTOFL	0	...	Dwg	Force line inside extension lines.
DIMTOH	1	...	Dwg	Text outside extensions is horizontal.
DIMTOL	0	...	Dwg	Generate dimension tolerances.
DIMTP	0.0000	...	Dwg	Plus tolerance.
DIMTSZ	0.0000	...	Dwg	Tick size.
DIMTVP	0.0000	...	Dwg	Text vertical position.
DIMTXT	0.1800	...	Dwg	Text height.
DIMTZIN	0	...	Dwg	Tolerance zero suppression.

Variable	Default	Ro	Loc	Meaning
CIRCLERAD	0.0000	...	...	Most-recent circle radius: 0 No default.
CLAYER	"0"	...	Dwg	Current layer name.
CMDACTIVE	1	R/o	...	Type of current command: 1 Regular command. 2 Transparent command. 4 Script file. 8 Dialogue box.
CMDDIA	1	...	Cfg	Plot command interface: 0 Command line prompts. 1 Dialogue box.
CMDECHO	*1*	...	...	*AutoLISP command display:* *0 No command echoing.* *1 Command echoing.*
CMDNAMES	"SETVAR"	R/o	...	Current command.
COORDS	1	...	Dwg	Coordinate display style: 0 Updated by screen picks. 1 Continuous display. 2 Polar display upon request.
CVPORT	2	...	Dwg	Current viewport number: 2 Minimum (default).

D

Variable	Default	Ro	Loc	Meaning
DATE	2448860.54043252	R/o	...	Date and fraction in Julian format.
DBGLISTALL	*0*	...	...	*Toggle of some sort.*
DBMOD	*0*	*R/o*	...	*Drawing modified in these areas:* *0 No modification made.* *1 Entity database.* *2 Symbol table.* *4 Database variable.* *8 Window.* *16 View.*
DIASTAT	1	R/o	...	User exited dialogue box by clicking on: 0 Cancel button. 1 OK button.

DIMENSION VARIABLES

Variable	Default	Ro	Loc	Meaning
DIMALT	0	...	Dwg	Alternate units selected.
DIMALTD	2	...	Dwg	Alternate unit decimal places.
DIMALTF	25.4000	...	Dwg	Alternate unit scale factor.
DIMAPOST	""	...	Dwg	Suffix for alternate text.
DIMASO	1	...	Dwg	Create associative dimensions.
DIMASZ	0.1800	...	Dwg	Arrow size.

Variable	Default	Ro	Loc	Meaning
ANGBASE	0	...	Dwg	Direction of zero degtees relative to UCS

ANGDIR	0	...	Dwg	Rotation of angles:
				  0 Clockwise.
				  1 Counterclockwise.

📷 APERTURE	10	...	Cfg	Object snap aperture in pixels:
				  1 Minimum size.
				  10 Default size.
				  50 Maximum size.

📷 AREA	0.0000	R/o	...	Area measured by **Area**, **List**, or **Dblist**.

ATTDIA	0	...	Dwg	Attribute entry interface:
				  0 Command-line prompts.
				  1 Dialogue box.

ATTMODE	1	...	Dwg	Display of attributes:
				  0 Off.
				  1 Normal.
				  2 On.

ATTREQ	1	...	Dwg	Attribute values during insertion are:
				  0 Default values.
				  1 Prompt for values.

AUNITS	0	...	Dwg	Mode of angular units:
				  0 Decimal degrees.
				  1 Degrees-minutes-seconds.
				  2 Grads.
				  3 Radians.
				  4 Surveyor's units.

AUPREC	0	...	Dwg	Decimals places displayed by angles.
AXISMODE	*0*	...	*Dwg*	*Obsolete system variable.*
AXISUNIT	*0.0000*	...	*Dwg*	*Obsolete system variable.*

B

Variable	Default	Ro	Loc	Meaning
BACKZ	0.0000	R/o	Dwg	Back clipping plane offset.

📷 BLIPMODE	1	...	Dwg	Display of blip marks:
				  0 Off.
				  1 On.

C

Variable	Default	Ro	Loc	Meaning
CDATE	19950105.15560660	R/o	...	Current date and time in YyyyMmDd.HhMmSsDd format.
CECOLOR	"BYLAYER"	...	Dwg	Current entity color.
CELTYPE	"BYLAYER"	...	Dwg	Current layer color.
CHAMFERA	0.5000	...	Dwg	First chamfer distance
CHAMFERB	0.5000	...	Dwg	Second chamfer distance.

System Variables

APPENDIX A

AutoCAD LT stores information about the current state of itself, the drawing, and the operating system in *system variables*. The variables help programmers (who often work with toolbar and DIESEL macros) determine the state of the AutoCAD system.

TIPS

■ You get a list of system variables at the 'Command:' prompt with the ? option of the *SetVar* command:

```
Command: setvar
Variable name or ?: ?
Variable(s) to list <*>: * [Enter]
```

■ The **SetVar** command lets you can change the value of most, but not all, variables.

■ *Italicized system variables* are not listed by the **SetVar** ? command nor in AutoCAD LT's *User's Guide*.

■ Some system variables have the same name as a command, such as **Area**; other variables do not work at the 'Command:' prompt. These are prefixed by the 📼 character.

■ **Default Value:** The table lists all known system variables, including undocumented variables, along with the default values as set in the Aclt.Dwg prototype drawing.

■ **Ro:** Some system variables cannot be changed by the user or by programming; these are labeled "R/O" (short for read-only) in the table below.

■ **Loc:** The value of system variables are located in a variety of places:
- **Aclt**   AutoCAD LT executable (hard-coded)
- **Cfg**   Aclt.Cfg or Aclt.Xmx files
- **Dwg**   Current drawing
- **...**   Not saved

Variable	Default	Ro	Loc	Meaning
_PKSER	*167-999999*	R/o	Aclt	Software package serial number.
_SERVER	*0*	R/o	Cfg	Network authorization code.
A				
ACLTPREFIX	"d:\ACLT\"	R/o	...	Path spec'd by ACAD environment var.
ACLTVER	"2.0"	R/o	...	AutoCAD LT version number.
AFLAGS	0	...	...	Attribute display code:
				0 No mode specified.
				1 Invisible.
				2 Constant.
				4 Verify.
				8 Preset.

Diesel Programming *cont'd*

Get stuff variables:

- **getvar**   Get system variable: **$(getvar,***variable***)**
  - Return the value of a system *variable*.
    For example, if system variable **FilletRad** holds the fillet radius of 0.5000, then:
    $(getvar,filletrad) results in 0.5000.
  - **GetVar** works with all of LT's system variables (see Appendix A).

- **getenv**   Get Aclt.Ini variable: **$(getenv,***variable***)**
  - Return the value of a *variable* in initialization file Aclt.Ini.
    For example, if variable **MenuName** is Aclt.Mnu, then:
    $(getenv,menuname) results in aclt.mnu
  - **GetVar** only works with AutoCAD LT's INI variables located in the **[AutoCAD LT General]** section of Aclt.Ini (see Appendix B).

- **edtime**   Returns the date and time: **$(edtime,***time,format***)**
  - Returns any *date and time* (not necessarily the current date and time) as specified by *format*.
    For example, to display current date and time in 24-hour format:
    (edtime,$(getvar,time),DD MON"," YYYY"," HH:MM)
    results in "18 Sep, 1995, 17:25"

  - The date and day format specifiers are:
D	9 (date).
DD	09 (date padded with zero).
DDD	Mon (abbreviated day).
DDDD	Monday.

  - The month and year format specifiers are:
M	9 (month).
MO	09 (month padded with zero).
MON	Sep (abbreviated month).
MONTH	September.
YY	95 (abbreviated year).
YYYY	1995.

  - The time format specifiers are:
H	4 (hour).
HH	04 (hour padded with zero).
MM	25 (minutes).
SS	38 (seconds).
MSEC	500 (milliseconds, 1/1000th of a second).
AM/PM	Display AM or PM, as appropriate.
am/pm	Display am or pm, as appropriate.
A/P	Display A or P, as appropriate.
a/p	Display a or p, as appropriate.

- **upper** Converts string to upper case: **$(upper,** *value***)**
  - Returns the string *value* changed to all uppercase letters:
  `$(upper,"quick")` results in QUICK.

String Functions:

- **index** Extract one element of a comma-separated series: **$(index,***value,string***)**
  - Finds and returns the *value* found in *string*:
    For example, if system variable **LastPoint** holds the x,y,z-coordinates of (44.3906,2.1774,0.0000), then the following **Index** expression returns the value of the y-coordinate (index = 1):
    `$(index,1,$(getvar,lastpoint))` results in 2.17741935
  - The index *value* begins with 0; the second item is #1.
  - The index *string* must consist of a single argument holding a series of comma-separated values, such as system variables **EntMax, GridUnit, InsBase, LastPoint, LimMax, ScreenSize, SnapBase, Target, UscOrg, ViewCtr,** and **VsMax.**

- **nth** Extract the nth of several elements: **$(nth,***value,n0,n1,...,n7***)**
  - Finds and returns the nth *value* found in elements *n0* through *n7*: if *value* = 0, then **Nth** returns the first element *n0*.
    `$(nth,2,45,1,4)` results in 4
  - The index *value* begins with 0; the second item is #1.

- **strlen** String length: **$(strlen,***string***)**
  - Returns the length (number of characters) of the *string*:
    `$(strlen,"This is toast")` results in 13.
  - The *string* can also be a number:
    `$(strlen,24515)` results in 5.
  - **StrLen** works with system variables in conjunction with **GetVar**.
    For example, system variable **AcltVer** contains "2.0":
    `$(strlen,$(getvar,acltver)` results in 3.

- **substr** Substring: **$(substr,***string,start,length***)**
  - Returns the substring (portion of the *string*), starting at character position *string* and continuing on for *length* characters; if *length* is not specified, all characters following the *start* position are returned:
    `$(substr,"This is toast",9)` results in toast.
  - The *string* can also be a number:
    `$(substr,24515,2,3)` results in 451.
  - **SubStr** works with system variables in conjunction with **GetVar**.
    For example, system variable **CDate** contains today's date and time as 19950825.20205319 (using the format YYYYmmdd.HHmmssdd):
    `$(substr,$(getvar,cdate),5,4)` results in 0825, the current month and date (August 25).

Diesel Programming *cont'd*

`^C^Ctilemode` — The start of the menu macro, which cancels any other command and executes the **TileMode** system variable.

`$M=$(-,1,$(getvar,tilemode))`
This is the remainder of the menu macro. It changes the value of system variable **TileMode** to either 1 or 0:

- `$M=` — Start of a Diesel macro within a menu macro.
- `$(-,` — Subtraction function.
- `1,` — Subtract 1 from the value of ...
- `$(getvar,tilemode))` — ...system variable **TileMode**.

Conversion functions:

- **angtos**   Convert number into angle format: **$(angtos,** *value, format, precision*)
    - Converts the *value* into an angle using the optional *format* and *precision*; if *format* and *precision* are left out, AngToS uses the values stored in system variables **AUnits** and **AuPrec**:
    `$(angtos,45,1,4)` results in 58d18'36"
    - Format values:
        - 0   Decimal degrees (0.0000)
        - 1   Degrees-minutes-seconds (00d00'00".0000)
        - 2   Grad (0.0000g)
        - 3   Radian (0.0000r)
        - 4   Surveyor units (N0'0.0000"E)
    - Precision range: 0 to 8 decimal places (default = 4).

- **rtos**   Format real number into units: **$(angtos,** *value, format, precision*)
    - Converts the *value* into an angle using the optional *format* and *precision*; if *format* and *precision* are left out, RToS uses the values stored in system variables **LUnits** and **LuPrec**:
    `$(rtos,45,1,4)` results in 4.5000E+01
    - Format values:
        - 0   Scientific: exponent notation (0.0000E+00)
        - 1   Decimal: the default (0.0000)
        - 2   Engineering: feet and decimal inches (0' - 0.0000")
        - 4   Architectural: feet and fractional inches (0' - 0/64")
        - 3   Fractional: unitless (0 0/64)
    - Precision range:
        0 to 8 for decimals (default = 4)
        0 to 1/256 for fractional (default = 1/64).

- **fix**   Converts real number into an integer: **$(fix,** *value*)
    - Truncates the decimal portion of *value*, a real number:
    `$(fix,45.14)` results in 45

- **eq**   Identity: **$(eq,** *n1***,** *n2***)**
  - Returns 1 if *n1* is identical to *n2*; otherwise returns 0.
- **!=**   Not equal to: **$(!=,** *n1***,** *n2***)**
  - Returns 1 if *n1* is not equal to *n2*; otherwise returns 0.
- **<**   Less than: **$(<,** *n1***,** *n2***)**
  - Returns 1 if *n1* is less than *n2*; otherwise returns 0:
    $(<,$(getvar,elevation),97) results in 1 if the value of system variable **Elevation** is less than 97.
- **<=**   Less than or equal to: **$(<=,** *n1***,** *n2***)**
  - Returns 1 if *n1* is less than or equal to *n2*; otherwise returns 0.
- **>**   Greater than: **$(<,** *n1***,** *n2***)**
  - Returns 1 if *n1* is greater than *n2*; otherwise returns 0.
- **>=**   Greater than or equal to: **$(>=,** *n1***,** *n2***)**
  - Returns 1 if *n1* is greater than or equal to *n2*; otherwise returns 0.
- **and**   Logical and: **$(and,** *n1***,** *n2***, ...,** *n9***)**
  - Returns the bitwise logical 'and' of *n1* through *n2*.
- **or**   Logical or: **$(or,** *n1***,** *n2***, ...,** *n9***)**
  - Returns the bitwise logical 'or' of *n1* through *n2*.
- **xor**   Logical xor: **$(and,** *n1***,** *n2***, ...,** *n9***)**
  - Returns the bitwise logical 'xor' (negative or) of *n1* through *n2*.
- **if**   If: **$(if,** *expression***,** *true***,** *false***)**
  - If *expression* is true, performs *true*; otherwise, performs *false*; the value of true can be taken as '1' and false as '0'. This can be used to check the value of toggle system variables, such as **TileMode**:

```
[$(if,$(getvar,tilemode),!.)/TTileMode]^C^Ctilemode $M=$(-,1,$(getvar,tilemode))
```

This complex-looking text is a menu macro that toggles the state of the TileMode system variable (switches the value between 0 and 1), as well as toggling the display of the checkmark in front of the TileMode label in the menu. When TileMode = 1, the checkmark appears. To help understand it, let's break down this menu and Diesel macro, bit by bit:

`[`	Start of the menu label.
`$(if,`	Start of the Diesel **If** function.
`$(getvar,tilemode),`	The *expression*, which obtains the value of variable **TileMode**.
`!.`	The *true*, which displays the checkmark, ✓, in front of the TileMode menu item if **TileMode** = 1.
`)`	The end of the Diesel expression; there is no *false* part.
`/TTileMode]`	The remainder of the menu label, <u>T</u>ileMode.

Diesel Programming *cont'd*

Diesel Functions

All Diesel functions take at least one variable; some take as many as nine variables.

Math functions:

- **+** Addition: $(+, *n1, n2, ..., n9*)
  - Add two or more numbers together:
    `$(+,1,3,5,7,11)` results in 27:

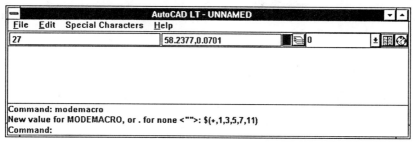

  - Add a number to a system variable:
    `$(+,96,$(getvar,elevation))` adds 96 inches (8 feet) to the current value of system variable **Elevation**.

- **-** Subtraction: $(-, *n1, n2, ..., n9*)
  - Subtract one or more numbers from the first:
    `$(-,11,3,5,7)` results in -4.

- ***** Multiplication: $(*, *n1, n2, ..., n9*)
  - Multiply two or more numbers:
    `$(*,3,5,7)` results in 105.
  - Multiply two or more numbers in scientific notation:
    `$(*,3e+5,7e+11)` results in 2.10000000e+17. (The largest displayable number is 1.0e+308.)

- **/** Division: $(/, *n1, n2, ..., n9*)
  - Divide one or more numbers into the first number:
    `$(/,13,5,7)` results in 0.37242857. (Results are displayed to eight decimal places.)

Logical functions:

- **=** Equality: $(=, *n1, n2*)
  - Returns 1 if two variables are equal; returns 0 if unequal:
    `$(=,$(getvar,elevation),96)` results in 1 if system variable **Elevation** = 96.
  - Check if a system variable is toggled on or off:
    `$(=,$(getvar,gridmode),1)` results in 1 if system variable **Gridmode** = 1 (grid is turned on).

Diesel Programming

Diesel programming was orginally introduced in AutoCAD Release 12 for DOS. Its purpose was to allow customization of the status line, which by default is the display of the O, S, and P indicators, the x,y-coordinates, the current color, and the layer name. Diesel allows users to change the status line to display other information, such as the time, the filename, and the z-coordinate. Diesel is the acronym for "direct interactively evaluated string expression language."

Autodesk chose not to include the powerful AutoLISP programming language in AutoCAD LT. The only available replacement in LT is the more cumbersome Diesel; hence, you see Diesel statements in menu macros that require decision making, such as toggling the display of the checkmark.

The default status line:

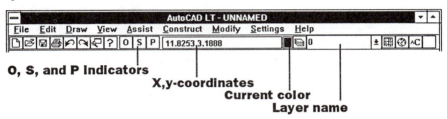

Diesel uses an unusual format for its macro language: $(*function*, *variable*), where *function* is one of 27 function names. The function acts on the variables, each separated by a comma. The function and variables are prefixed by '$(' and suffixed by ')'.

QUICK START: Diesel via ModeMacro

1. The easiest way to put Diesel to work modifying the status line is via the **ModeMacro** system variable:

    Command: **modemacro**
    New value for MODEMACRO, or . for none <"">: **The**
         Illustrated AutoCAD LT Quick Reference.

2. The area of the left of the coordinate readout is replaced by a text box containing part of the sentence:

The status line window can be used to display messages to the CAD operator. The text truncates after 32 characters, no matter how large the window.

6. Double-click on the DOS icon in the Program Manager:

7. Change to the \Acltwin subdirectory and run the **SlideLib** program ("parts" is the name of the slide library file, Parts.Slb):

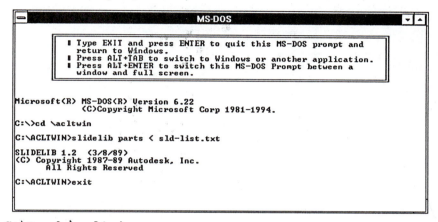

```
C:\> cd \acltwin
C:\acltwin\> slidelib parts < sld-list.txt
```

The SlideLib program creates the Parts.Slb slide library from the SLD slide files listed in the Sld-List.Txt file.

8. Exit from DOS back to Windows:
   C:\acltwin\> **exit**

9. Write the menu macro to display the icons:
   ***icon**
   ****icon_parts**
   [parts(cog,Six Segment Cog)]^C^Cinsert;cog;

10. Save the MNU file and load with the **Preferences** command.

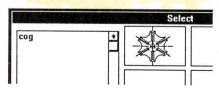

QUICK START: Converting Slides Into Icons

How to create slides for the icon menu:

1. To fit the area provided for the icon, the slide should have an aspect ratio of 2:3. Start AutoCAD with a new drawing and set up a viewport with the 2:3 aspect ratio. Use these commands:

        Tilemode 0
        MView 0,0 3,2
        Zoom Extents
        MSpace
        Fillmode 0

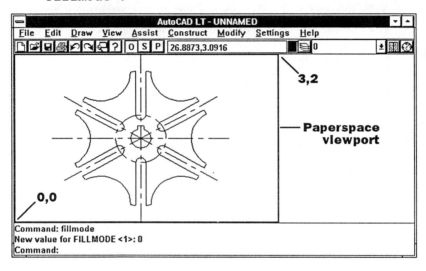

2. Draw the symbol or import a drawing with the **Insert** command.

3. After drawing the symbol in AutoCAD, save the image as an SLD slide file with the **MSlide** command.

4. Use the Notepad text editor to create the list of slide files to convert into a slide library file.

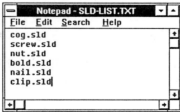

5. Save the file by any name with a TXT extension, such as Sld-List.Txt, in the \Acltwin subdirectory.

Menu Programming *cont'd*

QUICK START: Designing an Icon Menu

The icon menu looks a bit like a dialogue box, with a text list (at the left) and up to 16 "icon" image tiles on the right. The images are not icons (as on toolbar buttons); instead, they are small slides made with the **MSlide** command and **SlideLib.Exe** utility program. AutoCAD automatically generates the icon menu using just three macro programming elements:

Title bar: [Select Text Font]
Slide Image: [aclt(romans,
Label text: Roman Simplex)]

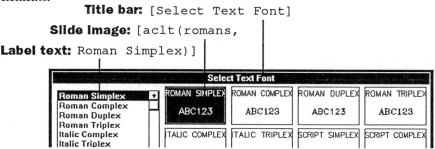

How to create an icon menu:

1. Open the MNU file and start with an *****icon** section:
    ***icon
This one icon section holds all icon menus.

2. Each separate icon menu starts with the ** (submenu section) tag. Start the icon menu with an ********icon_name*, such as:
    **icon_fonts1

3. For the icon menu's title bar, place a label in square brackets, such as:
    [Select Text Font]

4. Finally, a list of image tiles (shown in boldface) and related macro:
    [aclt(romans,Roman Simplex)]'_style romans romans

This icon menu macro, taken from the Aclt.Mnu file, helps you select a text style. When you click on the picture of the RomanS font or select "Roman Simplex" from the list, AutoCAD runs the **Style** command. Here's what the elements mean:

- **[aclt**   Name of the slide library file, Aclt.Slb.
- **(romans,**   Name of the slide image, the icon.
- **Roman Simplex)]**   The label text.
- **'_style**   Start the macro with the **Style** command.
- **romans romans**   Answer the first two **Style** command prompts with the name of the text style, RomanS.

3. Clicking on ✓**Associative Dimensions** turns the check mark on and off:

Settings	Settings
✓Associati̱ve Dimensions	Associati̱ve Dimensions

■ !.   Prefix label with a check mark; for example:
!./vAssociative Dimensions

To display the checkmark is easy but turning the check mark on and off is tricky and requires the use of Diesel programming.

[$(if,$(getvar,dimaso**),!.)/vAssociative Dimensions]'_dimaso $m=$(-,1,$(getvar,**dimaso**))**

The code shown above is a combination of menu macro code and Diesel programming code. The code checks whether system variable **DimAso** is 1 (on = show check mark) or 0 (off = don't display check mark). For more information, see the chapter on *Programming Diesel*.

The quick-start workaround is to copy the elements shown in **boldface**. Replace the parts in *italic* with the appropriate text and system variables. For example, here is the same code for toggling the grid:

[$(if,$(getvar,gridmode**),!.)/GGrid Markings]'_gridmode $m=$(-,1,$(getvar,**gridmode**))**

(The macro is one long line; don't split it into two lines in your menu file!)

Menu Programming *cont'd*

QUICK START: Advanced Menu Tricks

How to create child and sub menus, and toggle a check mark:

1. To create the child menu:
```
[--]
[->/IImport/Export]
  [/VView Slide...]^C^C_vslide
  ...
  [<-/PBMP Out...]^C^C_savedib
```
Clicking on the **Import/Export** ▶ menu item displays the child menu:

 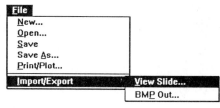

- ▪ ->   Adds the ▶ character and triggers the child menu.
- ▪ <-   Indicates the last item of the child menu.
- ▪ [--]   Draws a horizontal line between menu items.

2. Related to the child menu is the submenu. This is where AutoCAD branches off to another part of the menu file. Aclt.Mnu contains just one submenu:
```
...
[/TText Style...]$I=icon_fonts1 $I=*
...
***icon
**icon_fonts1
[Select Text Font]
[aclt(romans,Roman Simplex)]'_style romans romans
```
The boldface text, above, creates the jump using these elements
- ▪ *$I=icon_fonts1*   Jump to an icon submenu named "icon_fonts1":
  - ▪ $ Jump.
  - ▪ I   Type of submenu:
    - ▪ P0   Cursor menu.
    - ▪ P*n*   Menu bar item *n* = 1 to 16.
    - ▪ A*n*   Mouse button menu *n* = 1 to 4.
    - ▪ I   Icon menu.
  - ▪ =   Name of the submenu, "icon_fonts1" in this case.
- ▪ ***icon_fonts1*   Start of the submenu section.
- ▪ $I=*   Displays the currently loaded menu; in this case, **icont_fonts1.
- ▪ $I=   Returns to the previous menu.

- [~*label*]   Disable the menu item: text is displayed in grey and macro does not run; child menus are also disabled. Use the **boldface** portions of the following Diesel macro to toggle the disabled/enabled state:
  [$(if,$(getvar,tilemode**),~)**Paper + Space]^C^C^C_pspace;
- [!c *label*]   Prefix label with a checkmark.
- !n   Display a special character. Normally, special characters like [, $, and ~ are not displayed because they perform macro functions. Prefixing the special character causes it to be displayed, rather than executed:

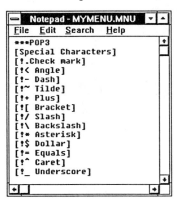
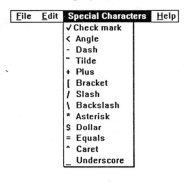

Menu Programming *cont'd*

- **$M=($code)** Include Diesel code in macro, as shown in **boldface** below:
    **[$(if,$(getvar,dimaso),!.)/vAssociative +
    Dimensions]'_dimaso $m=$(-, 1, $(getvar, +
    dimaso))**

3. Control characters:
    - **^B** Toggle (turn on and off) snap mode, like pressing **[Ctrl]+B** or using **\2** in a toolbar macro; '**^**' is short for **[Ctrl]**; for example:
        [Snap Toggle]'^B
    - **^C** Cancel the current command; for example:
        [Insert Block]^C^C^C_insert;\\\\
    - **^D** Toggle the real-time display of coordinates.
    - **^G** Toggle grid display.
    - **^H** Backspace to suppress the automatic **[Enter]**; for example selecting angle from a menu (the 'x' is a sacrificial character):
        [90]90x^H
    - **^O** Toggle ortho mode.
    - **^P** Toggle menu echoing; by default, the menu macro is displayed at the 'Command:' prompt. Adding the ^P to the macro turns off the echoing:
        [Insert Block]^P^C^C^C_insert;\\\\
    - **^V** Switch focus to the next viewport.
    - **^Z** Suppress the **[Enter]** at the end of the macro.
    - **^@** Null character (same as ^Z).
    - **^[** Escape character, like ASCII code 27.
    - **^** File separator.
    - **^]** Group separator.
    - **^^** Record separator.
    - **^_** Unit separator.

4. Lable Characters
    - **[*label*]** Start of the macro; for example:
        [Insert Block]insert
    - **[- -]** Draws a separator line in pull-down and cursor menus.
    - **[-> *label*]** Starts a child menu.
    - **[<- *label*]** End of the child menu.
    - **[<-<- *label*]** End of child and parent menu.

QUICK START: Menu Macro Syntax

Here are all the special characters that LT menu macros understand:

1. Section characters:
- ***label**   Start of section; for example: `***POP1`
  Valid section names are:
  - POP0   Cursor menu.
  - POP1 *thru* POP16   Menu bar.
  - AUX1 *thru* AUX4   Mouse button menus.
  - ICON   Icon menu.

- ****label** Start a subsection; for example: `**File`
  You can use anything for the label.

- **$n=label** Jump to a subsection; AutoCAD remembers a maximum of 8 jumps for the return paths; for example: `$I=icon_fonts1`
  Valid jump-to section names are:
  - P0   Cursor menu.
  - P1 *thru* P16   Menu bar.
  - A1 *thru* A4   Mouse button menus.
  - I   Icon menu.

- **$n=*** Display the current subsection; for example: `$P1=*`
- **$n=** Return to previous subsection; for example: `$A1=`

2. Macro characters:
- ****   (Backslash) Pause for user input; macro continues after you pick a point or press **[Enter]** or **[Spacebar]**; for example:
  `[Insert Block]insert \\\\`

- **;**   (Semi-colon) Equivalent of pressing **[Enter]** and easier to see than the space character in macros; for example:
  `[Insert Block]insert;\\\\`

- **_**   (Underscore) Internationalize the command; for example:
  `[Insert Block]_insert;\\\\`

- ***** (Asterisk) Repeat the command until canceled with **[Ctrl]+C**
  (* is equivalent to the **Multiple** command); for example:
  `[Insert Block]*_insert;\\\\`

- **'**   (Apostrophe) Transparent command (execute the command within another command); for example:
  `[Zoom Windows]'_zoom w \\`

- **+**   (Plus) Continue a macro on the next line (useful for very long macros); see the + signs in the example below.

Menu Programming *cont'd*

QUICK START: Reloading a Menu File
How to load the menu file into AutoCAD after making a change:

1. Start a text editor, such as Notepad.

2. Load the Aclt.Mnu file, which contains the source code for the menus:

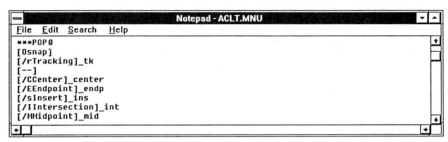

4. Make your editing changes.

5. Save the menu file by another name, such as MyMenu.Mnu.

6. Load the MyMenu.Mnu file with the **Preferences** command.

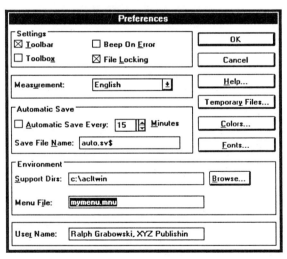

7. When the **Preferences** dialogue box appears, type:
   Menu File: **mymenu.mnu**

8. Click on the **OK** button and AutoCAD loads the changed menu file. (The **Menu** command documented by the *LT User's Guide* does not exist in AutoCAD LT.)

9. AutoCAD LT does not reload the menu file unless the name is changed. The next time you want to load an edited MNU file, give it a slightly different name, such as MyMenu1.Mnu.

QUICK START: Creating a Simple Menu

How to create a custom menu:

1. Start the Notepad editor.

2. Decide what kind of menu you want to create — pulldown menu, cursor menu, button menu, or icon menu — then write the label, such as:
```
***POP1
```

3. Give the menu item a name using these characters:
- [ ]   (Square brackets) Name the menu item.
- /n   Underline a character for the [Alt]+n keyboard shortcut.

For example:
```
***POP1
[/FFile]
```
Creates the **File** item on the menu bar:

4. To have a menu item execute an AutoCAD command, type the command at the appropriate spot in the Aclt.Mnu file. For example, here is the **New** command added under the **File** item on the menu bar:
```
***POP1
[/FFile]
[/NNew...]^C^C_new
```

5. The menu macro has the following syntax:
- [Start of label.
- /N   Underline the N.
- New   Label name.
- ...   Alert user that this will open a dialogue box.
-]   End of label.
- ^C^C   As with toolbar macros, it is prudent to start commands with a pair of cancels. Instead of \3\3, menu macros use ^C^C.
- _   Internationalize the command.
- New   The **New** command (finally!).

6. Save your menu file and load into AutoCAD with the **Preferences** command.

Menu Programming *cont'd*

TIPS

■ Each pull-down menu (POP*n*) can have a maximum of 999 items including all child menu items.

■ The cursor menu (POP0) can have a maximum of 499 items.

■ The button menu (AUX*n*) can have a maximum of 15 items; the meaning of the first button is always **[pick]** and cannot be changed.

■ The resolution of the screen provides a more practical limit, such as 20 items.

■ Interface designers use the rule of "five items, plus or minus two": a minimum of 3, a maximum of 7.

■ The cursor menu (POP0) does not operate unless at least one pull-down menu (POP1) is defined.

■ The pulldown and cursor menus are made automatically wide enough to fit the widest text label.

■ The longest menu bar title is 14 characters.

■ The cursor menu does not display a title bar, unlike the pull-down and icon menus, but still must have a dummy section title.

■ AutoCAD LT forces the **File**, **Edit**, and **Help** items on the menu bar, along with these Windows-specific menu items:

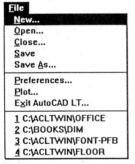 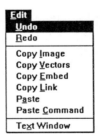 

■ If you create menu sections with titles **[File]** and **[Edit]**, LT automatically adds in the above Windows-specific menu items.

■ If you don't have any sections named **[File]** or **[Edit]**, AutoCAD adds the menus shown above. All your menus appear after these first two.

- Button #1 (leftmost button) is always the [pick] button and cannot be redefined.
- On a two-button mouse, press [Shift]+[right button] to display the screen menu.
- The AUX*n* sections can be used to define additional functions for the mouse buttons in combination with the [Ctrl], [Shift], and [Ctrl]+[Shift] keys.
- The inclusion of ^T is odd since AutoCAD LT does not support the tablet toggle (normally [Ctrl]+T).

Understanding the Menu Structure

Load the Aclt.Mnu file into the Notepad text editor and take a look at the menu file. The three different menus (menu bar, cursor menu, and mouse button menu) are defined by section names prefixed by *** (three asterisks):

- ***Pop0   The cursor menu.
- ***Pop*n*   The menu bar; maximum = 16 items; the Aclt.Mnu file written by Autodesk defines the following items on the menu bar:

  File Edit Draw View Assist Construct Modify Settings Help

  - ***Pop1   File
  - ***Pop3   Draw
  - ***Pop4   View
  - ***Pop5   Assist
  - ***Pop6   Construct
  - ***Pop7   Modify
  - ***Pop8   Settings

The **File, Edit,** and **Help** items *cannot* be removed. You can add menu items ***Pop2 and ***Pop9 through ***Pop16.

- ***Aux1   The mouse buttons.
- ***Aux*n*   Mouse button (🖱) in combination with [Shift] and [Ctrl] keys:
  - ***Aux2   [Shift]+🖱
  - ***Aux3   [Ctrl]+🖱
  - ***Aux4   [Ctrl]+[Shift]+🖱

- ***Icon   Icon menu
- ***Comment   Comment section ignored by AutoCAD.

Menu Programming

The menu bar, pop-up cursor menu, and mouse buttons are preconfigured by Autodesk. You can change the command assigned to almost any menu item; as well, you can add more menu items, move the position of menu items, and change the wording of menu items. You *cannot* change the **Edit** and **Help** menus.

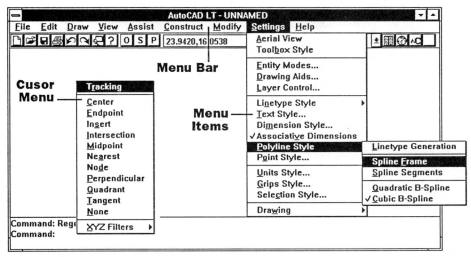

Menu items use the following notations:

- _ (Underscore) Keyboard shortcut, such as **Drawing** is accessed by pressing **[Alt]+S W**
- ... (Ellipses) Displays a dialogue box, such as **Entity Modes...** displays the **Entity Creation Modes** dialogue box.
- ✓ (Check mark) Item is toggled on, such as ✓**Associative Dimensions**.
- ▶ (Arrowhead) Displays a child menu, such as **Polyline Style** ▶

To display the screen menu, click the middle mouse button (or **[Shift]+[Right button]**).

Mouse Buttons

By default, the mouse buttons have the following meaning (assuming you have a mouse with more than three buttons — there are some on the market):

Button	Macro	Meaning	Notes
2	;	[Enter]	The rightmost button.
3	$p0=*	Display screen menu.	The middle button.
4	^C^C	Cancel command.	
5	^B	Snap toggle.	
6	^O	Ortho toggle.	
7	^G	Grid toggle.	
8	^D	Coordinate toggle.	
9	^E	Isoplane toggle.	
10	^T	Table toggle.	Not valid in AutoCAD LT.

QUICK START: Writing a Macro

How to string together more than one command — this is called a "macro." The **Toolbar Customization** dialogue box, below, shows a single button assigned three consecutive commands:

    `AutoCAD LT Command: zoom e save plot`

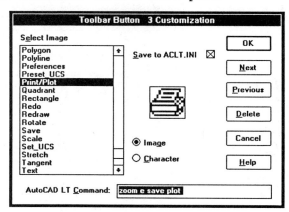

These replace the **Plot** button with a useful three-step macro that zooms and saves the drawing before plotting:
1. **Zoom E**   Zooms the drawing to its extents.
2. **Save**   Saves the drawing.
3. **Plot**   Displays the **Plot Configuration** dialogue box.

With a single click of the button, this macro saves 17 keystrokes or seven menu picks. To access this dialogue box, click the rightmost mouse button over the toolbar or toolbox you want to change.

Toolbar Programming *cont'd*

QUICK START: Changing a Toolbar/Toolbox Button

How to make any change to a toolbar or toolbox button:

1. Move the cursor over the button.

2. Press the rightmost button on the mouse. AutoCAD displays the **Toolbar Button Customization** dialogue box:

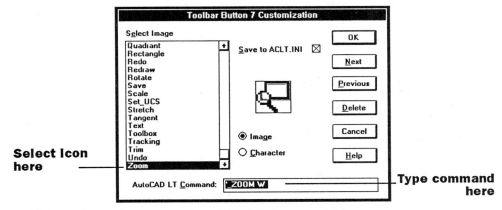

3. Click on the text box next to **AutoCAD LT Command**.

4. Erase the existing command and type the new command.

5. The icon is the small picture that labels the button. To change the icon, select a different name from the list under **Select Image**. (The name of the icon becomes its tooltip label.) Only toolbar icons can have a letter of the alphabet, rather than an icon.

6. Click on the **OK** button to dismiss the dialogue box.

7. Click on the redefined button to check that it works correctly.

- **** Allows use of the \ (backslash) character in macros, such subdirectory.
  For example: `fileopen c:\\acltwin\\office.dwg` results in
  ```
  Command: fileopen
  Enter name of drawing: c:\acltwin\office.dwg
  ```

TIPS

- A toolbar and toolbox macro can be at most 255 characters long.

- If you need to write a macro longer than 255 characters, use command aliases to save on space (see Appendix C). For example, using aliases for 'zoom e save plot' results in saving 50% of characters:
  ```
  AutoCAD LT Command: z e sa pp
  ```

- The maximum number of macros (buttons) that LT can display at one time:
  - **Toolbar**   26 (11, at 640x480 resolution)
  - **Toolbox**   60
  - **Total**   86

- If you need more than 86 macros, consider storing the additional macros in a separate Aclt.Ini file kept in another subdirectory.

- Macros are stored in the Aclt.Ini file:
  - **Toolbar**   [AutoCAD LT General] section
  - **Toolbox**   [AutoCAD LT ToolBox] section

- You can edit the macros directly by editing the Aclt.Ini file with Notepad.

- Always start the macro with **\3\3\3**. This ensures that any command currently underway is cancelled. While Autodesk recommends just **\3\3**, the extra **\3** ensures cancellation of "deep" commands, such as **PEdit** and **Dim**. For example:
  ```
  AutoCAD LT Command: \3\3\3line
  ```

- The only exception to using the **\3\3\3** prefix is when you want the macro to operate within another command. For example, to zoom to the previous view during another command:
  ```
  AutoCAD LT Command: 'zoom p
  ```

- Toolbar-toolbox macros cannot pause for user input. To get around this, write two macros: one for before the user input, the other for after user input. For example, to supply the coordinates of a block being repeatedly inserted:
  - **Click button 1:**   \3\3\3insert
  - **[Pick] insertion point.**
  - **Click button 2:**   1 1 0

Toolbar Programming *cont'd*

- **\3** Cancel any command currently underway, like pressing **[Ctrl]+C**; useful for ensuring that the macro starts at the 'Command:' prompt.
  For example: `\3\3line` results in:
  `Command: *Cancel* line From point:`

- **\4** Toggle coordinate display between on and off, like **[Ctrl]+D**. Does not toggle relative display mode.
  For example: `line\D` results in:
  `Command: line <Coords off> From point:`

- **\5** Switch between the three isometric planes, like **[Ctrl]+E**. Toggles from right to left to top isoplane.
  For example: `\5ellipse` results in:
  `Command: <Isoplane top> ellipse`

- **\7** Toggle display of the grid between on and off, like **[Ctrl]+G**.
  For example: `\7insert` results in:
  `Command: <Grid on> insert`

- **\8** Backspace, like pressing the **[Backspace]** key.
  Not documented by Autodesk.

- **\13** Enter, like pressing the **[Enter]** key; required during **DText** command.
  Not documented by Autodesk.
  For example: `dtext 2,2 0.5 0 First line\13Second line\13\13`; results in:
  ```
  Command: dext
  Justify/Style/<Start point>: 2,2
  Height: 0.5
  Rotation angle: 0
  Text: First line
  Text: Second line
  Text:
  Command:
  ```

- **\15** Toggle ortho mode between on and off, like **[Ctrl]+O**.
  Not documented by Autodesk.

- **\22** Switch to next viewport, like **[Ctrl]+V**.
  Not documented by Autodesk.
  For example: `\22zoom w` results in:
  `Command: zoom w (focus is switched to next viewport).`

- **\n** Start a new line; does *not* work with **DText** command.

- **\t** Tab; same effect as pressing **[Spacebar]**. Undocumented by Autodesk.

- **\nnn** Any ASCII character *nnn* from 000 to 255. Undocumented by Autodesk. *For example*: to toggle the grid, `\007` is the same as `\7`.

Toolbar Programming

The toolbar and toolbox buttons are preconfigured by Autodesk with 60 commands. You can change the icon and the command assigned to any button; as well, you can add more buttons (to a maximum of 86).

You cannot move the position of buttons, change the wording of tooltips, change the **O**, **S**, or **P** buttons, nor create new icons.

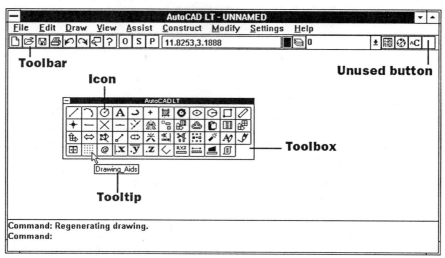

Special Characters

In addition to command and option names, you can use a limited number of special characters. For example, the space between each command is interpreted by LT as the same as pressing the [Enter] key. The complete list of special characters is as follows:

- **[Spacebar]** Equivalent to the user pressing [Enter] at the end of a command name.
  For example: **zoom e** results in:
   Command: zoom extents Regenerating drawing.

- **;** (Semi-colon) LT automatically adds a space at the end of the macro; adding the semi-colon suppresses the space.
  For example: **.y;**

- **\2** Toggle snap mode, just like pressing [Ctrl]+B on the keyboard; "toggle" means to turn on if off, and turn off if on.
  For example: **\2line** results in:
   Command: <Snap on> line From point:

Note the correlation between the letter 'B' (as in [Ctrl]+B being the second letter of the alphabet) and the '2' in the special character \2.

3dPoly

Rel. 1

Draws 3D polylines (*short for 3D POLYline*).

Command	Alt+	Menu bar	Alias	GC Alias
3dpoly	...	...	...	...

```
Command: 3dpoly
From point: [pick]
Close/Undo/<Endpoint of line>: [pick]
```

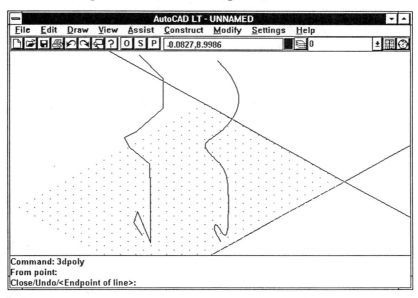

COMMAND OPTIONS

Close   Join the last endpoint with the start point.
<Endpoint of line>
              Indicate the endpoint of the current segment.
Undo   Erase the last-drawn segment.

RELATED AUTOCAD COMMANDS
- **Explode**   Reduces a 3D polyline into lines and arcs.
- **PEdit**   Edits 3D polylines.
- **PLine**   Draws 2D polylines.

TIPS
- Since 3D polylines are made of straight lines, use the **PEdit** command to spline the polyline as a curve.

- 3D polylines do not support linetypes and widths.

RELATED SYSTEM VARIABLES
- **ViewCtr**   Coordinates of the current view's center point.
- **ViewSize** Height of the current view.
- **VsMax**   Upper-right corner of the virtual screen.
- **VsMin**   Lower-left corner of the virtual screen.

TIPS
- A scale factor of one displays the entire drawing as defined by the limits.

- A zoom factor of:
    - 2   Enlarges objects (zooms in).
    - 1   Does not change the zoom level.
    - 0.5   Makes objects smaller (zooms out).

- The **Zoom Previous** option remembers the ten previous view settings, whether executed by the **Pan, View,** or **Zoom** command.

- The default for the **Zoom** command is the **Window** option, not the **Scale** option indicated by the <default> angle brackets.

 # 'Zoom

Rel. 1

Displays a drawing larger or smaller in the current viewport.

Command	Alt+	Menu bar	Alias	GC Aliases
'zoom	V,Z	[View]	z	za, zi, zl
		[Zoom]		zm, zo, zp
				zv, zw

Command: **zoom**
All/Center/Extents/Previous/Window/<Scale(X/XP)>: [Pick]
Other corner: [pick]

Before Zoom Window: *After Zoom Window:*

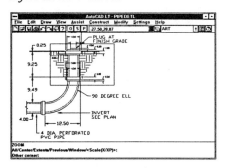 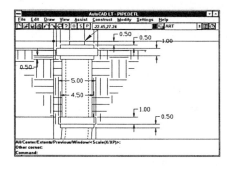

COMMAND OPTIONS

All   Displays the drawing limits or extents, whichever is greater.
Center   Zooms into a center point:
   Center point   Indicate the center point of the new view.
   Magnification or Height
                Indicate a magnification value or height of view.

Extents   Displays the current drawing extents.
Previous   Displays the previous view generated by **Pan**, **View**, or **Zoom**.
Window   Indicate the two corners of the new view (the default).
<Scale(X/XP)> Display a new view as a factor of the drawing limits:
   X   Display a new view as a factor of the current view.
   XP   Display a paper space view as a factor of model space.

[pick]   Begin **Window** option.

RELATED AUTOCAD COMMANDS
- **DsViewer**   Aerial View window performs zooms and pans.
- **Limits**   Specifies the limits of the drawing.
- **Pan**   Moves the view to a different location.
- **View**   Saves views by name.

Xref

Rel. 1

Attaches a drawing to the current drawing (*short for eXternal REFerence*).

Command	Alt+	Menu bar	Alias	GC Alias
xref	D,X	[Draw] [External Reference]	xr	...

Command: **xref**
?/Bind/Detach/Path/Reload/<Attach>:

COMMAND OPTIONS
<Attach>	Attach another drawing to the current drawing.
Bind	Make the externally referenced drawing part of the current drawing.
Detach	Remove the externally referenced drawing.
Path	Respecify the path to the externally referenced drawing.
Reload	Update the externally referenced drawing.
?	List the names of externally referenced drawings.

RELATED AUTOCAD COMMANDS
- **Insert**   Adds another drawing to the current drawing.
- **XBind**   Binds parts of the externally referenced drawing to the current drawing.

TIPS
■ The **XRef** command lets you view other drawings at the same time as the currently loaded drawing; however, you cannot edit the externally referenced drawing.

■ When you load a drawing that contains xrefs, the xref'ed drawings are also automatically loaded.

■ Using xrefs is a good way to work with very large drawings, since it reduces the amount of memory AutoCAD needs.

■ If you are working with AutoCAD on a network, no other user can access the externally referenced drawing while you are loading it; this is called "soft file locking."

XBind

Rel. 1

Binds portions of an externally referenced drawing to the current drawing (*short for eXternal BINDing*).

Command	Alt+	Menu bar	Alias	GC Alias
xbind	D,X,M	[Draw] [External Reference] [Bind Symbols]	xb	...

```
Command: xbind
Block/Dimstyle/LAyer/LType/Style: b
Dependent Block name(s):
```

COMMAND OPTIONS

Block	Binds blocks to current drawing.
Dimstyle	Binds dimension styles to current drawing.
LAyer	Binds layer names to current drawing.
LType	Binds linetype definitions to current drawing.
Style	Binds text styles to current drawing.

RELATED AUTOCAD COMMANDS
- **Insert**   Inserts a DWG file from the hard drive into the drawing as a block.
- **XRef**   Attaches another drawing to the current drawing.

RELATED SYSTEM VARIABLES
- *None.*

WmfOut

Rel. 1

Exports a WMF vector file (*short for Windows MetaFile OUTput*).

Command	Alt+	Menu bar	Alias	GC Alias
wmfout	F,I,M	[File] [Import/Export] [WMF Out]	wo	...

Command: **wmfout**
Displays dialogue box:

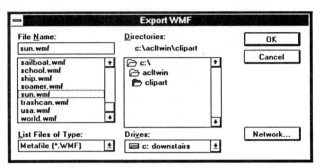

```
Select objects: [pick]
Select objects: [Enter]
```

COMMAND OPTIONS

OK	Accept filename.
Cancel	Dismiss the dialogue box and cancel command.
Network	Displays the **Connect Network Drive** dialogue box.
Select objects:	Select objects to be included in WMF file using any selection method.

RELATED AUTOCAD COMMANDS

- **CopyClip** Copies objects to the Windows Clipboard in picture (WMF) format.
- **CopyEmbed** Copies objects to the Clipboard in picture format for OLE embedding in another application.
- **CopyLink** Copies objects to the Clipboard in picture format for OLE linking in another application.
- **WmfIn**   Imports files in WMF format.

TIPS

- WMF files created by AutoCAD are resolution-dependent; small circles and arcs loose their roundness.

- The 'All' selection does not select all objects in drawing; instead, **WmfOut** selects all objects visible in the current viewport.

WmfOpts

Rel. 1

Controls the importation of WMF files.

Command	Alt+	Menu bar	Alias	GC Alias
wmfin	F,I,O	[File] [Import/Export] [WMF In Options]	...	...

Command: **wmfopts**
Displays dialogue box:

COMMAND OPTIONS

Wireframe   On: No area fills (default).
                 Off: Filled areas are imported as solids.
Wide lines   On: Lines retain their width (default).
                 Off: Lines are collapsed to zero width.

RELATED AUTOCAD COMMANDS

- **WmfIn**   Imports WMF files.
- **WmfOut**   Exports selected objects in WMF format.

Y scale factor The scale in the y-direction (default = x-scale factor).
Rotation angle The angle to rotate the WMF picture counterclockwise from 0 degrees.

RELATED AUTOCAD COMMANDS
- **PasteClip** Imports WMF files (picture) from the Windows Clipboard.
- **WmfOpts** Controls the importation of WMF files.
- **WmfOut** Exports selected objects in WMF format.

RELATED FILE
- AutoCAD LT Release 2 includes 39 WMF files in subdirectory \Acltwin\Clipart.

TIPS
- The WMF is placed as a block with the name WMF0; subsequent placements increment the digit: WMF1, WMF2, et cetera.

- Exploding the WMF*n* block results in polylines; even circles, arcs, and text are converted to polylines; solid-filled areas are exploded into solid triangles.

- When you import a CLP (short for CLiPboard) file with the **WmfIn** command, AutoCAD rejects all non-vector data:

WmfIn

Rel. 1

Imports CLP and WMF vector (picture) files (*short for Windows MetaFile INput*).

Command	Alt+	Menu bar	Alias	GC Alias
wmfin	F,I,W	[File] [Import/Export] [WMF In]	wi	...

Command: **wmfin**
Displays dialogue box:

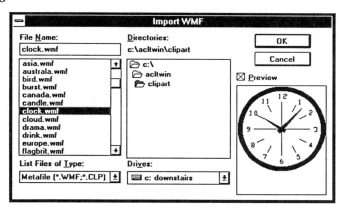

*After you select the filename, AutoCAD converts the CLP or WMF file into a block and continues with the **Insert** command:*

```
_INSERT Block name (or ?): WMF0
Insertion point:    [pick]
X scale factor <1>/Corner/XYZ: [Enter]
Y scale factor (default = X): [Enter]
Rotation angle <0>: [Enter]
```

COMMAND OPTIONS

OK   Select WMF filename and place in drawing.
Cancel   Dismiss dialogue box and cancel command.
Preview   Display selected WMF file in Preview window.
Insertion point The x,y,z-coordinates of the insertion point.
X scale factor   The scale in the x-direction (default = 1.0).
  Corner   Show x,y-scale factors by picking two corners on screen.
  XYZ   Enter the x,y,z-scale factors.

WBlock

Rel. 1

Writes a block, or part, or all of the drawing to disk (*short for Write BLOCK*).

Command	Alt+	Menu bar	Alias	GC Alias
wblock	F,I,B	[File] [Import/Export] [Block Out]	w	sa

Command: **wblock**
File name:
Block name:

COMMAND OPTIONS

=	(Equals) Block is written to disk using block's name as filename.
*	(Asterisk) Entire drawing is written to disk.
[Enter]	Creates a block on disk of selected entities.
[Space]	Moves selected objects to the specified drawing.

RELATED AUTOCAD COMMANDS

- **Block**   Creates a block of a group of entities.
- **Insert**   Inserts DWG files into current drawing as a block.
- **XBind**   Binds in blocks from other drawings.

TIP

- The **WBlock** command exports blocks from the current drawing to individual DWG files on disk.

- The **WBlock** command is an alternative to the **Purge** command for ridding the drawing of unnecessary elements and reduce filesize. To do so, use the * option to write the entire drawing to disk using the drawing's filename; then use the **Quit** command.

VSlide *cont'd*

TIPS

- The SLD file created by AutoCAD is in a proprietary raster format and hence resolution dependent. For best viewing results, use the **MSlide** command with the AutoCAD LT graphics window maximized and a single viewport.

- For faster viewing of a series of slides during a script file, an asterisk preceeding **VSlide** preloads the slide file, as in:
    Command: ***vslide filename**

RELATED AUTOCAD COMMANDS
- MSlide   Creates an SLD-format slide file of the current viewport.
- Redraw   Erases the slide from the screen.

RELATED AUTODESK PROGRAM
- SlideLib.Exe A DOS-based program that creates an SLB-format library file of a group of slide files; used for creating icon menus.

RELATED SYSTEM VARIABLES
- *None.*

RELATED FILE
- ColorWh.Sld

  Displays AutoCAD's 255 colors in color wheel fashion by number and color group:

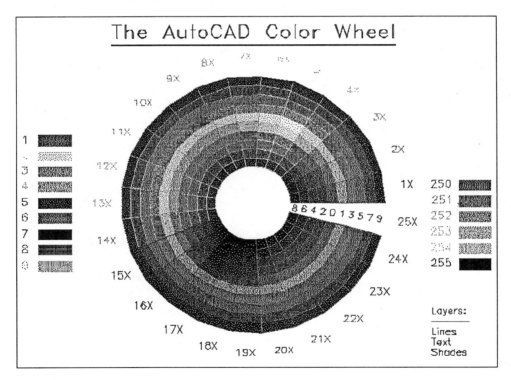

VSlide

Rel. 1

Displays an SLD-format slide file in the current viewport (*short for View SLIDE*).

Command	Alt+	Menu bar	Alias	GC Alias
vslide	F,I,S	[File] [Import/Export] [View Slide]	vs	...

Command: **vslide**
Displays dialogue box:

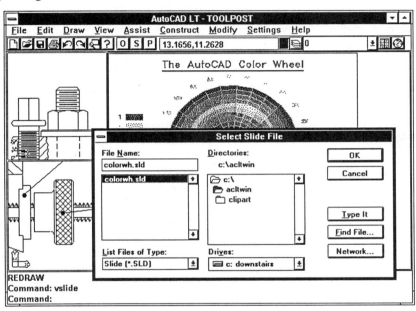

COMMAND OPTIONS

OK   Accept filename and display slide file.
Cancel   Dismiss dialogue box and cancel command.
Type It   Dismisses dialogue box and returns to the 'Command:' prompt:
                     Slide file: <filename>:

Find File   Displays **Find File** dialogue box to search for files on hard drive.
Network   Displays **Connect Network Drive** dialogue box.

<3>	Split the current viewport into three:
Horizontal	Create three viewports over each other.
Vertical	Create three viewports beside each other.
Above	Create two viewports over one viewport.
Below	Create two viewports below one viewport.
Left	Create two viewports left of one viewport.
<Right>	Create two viewports right of one viewport.
4	Split the current viewport into four.
?	List the names of saved viewport configurations.

RELATED AUTOCAD COMMANDS
- MView   Creates viewports in paper space.
- [Ctrl]+V   Moves focus to the next viewport.

RELATED SYSTEM VARIABLES
- CvPort   The current viewport.
- MaxActVp   The maximum number of active viewports.
- TileMode   Controls whether viewports can be overlapping or tiled.

VPorts or ViewPorts   Rel. 1

Creates viewports (or windows) of the current drawing: tiled when **TileMode** is on; overlapping when **TileMode** is off.

Commands	Alt+	Menu bar	Aliases	GC Alias
vports	V,O	[View] [Viewports]	vw	vp
viewports			[Ctrl]+V	

Command: **vports**
Save/Restore/Delete/Join/SIngle/?/2/<3>/4:

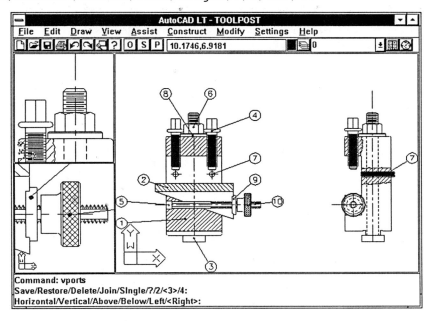

COMMAND OPTIONS
Delete	Delete a viewport definition.
Join	Join two viewports together as one.
Restore	Restore a viewport definition.
Save	Save the settings of a viewport by name.
SIngle	Join all viewports into a single viewport.
2	Split the current viewport into two:
Horizontal	Create one viewport over another.
<Vertical>	Create one viewport beside another.

VPoint

Rel. 1

Changes the viewpoint of a 3D drawing *(short for ViewPOINT)*.

Command	Alt+	Menu bar	Alias	GC Alias
vpoint	V,D	[View] [3D Viewpoint]	...	...
	V,W	[View] [3D Viewpoint Presets]		

Command: **vpoint**
Rotate/<View point> <0.0000,0.0000,1.0000>:

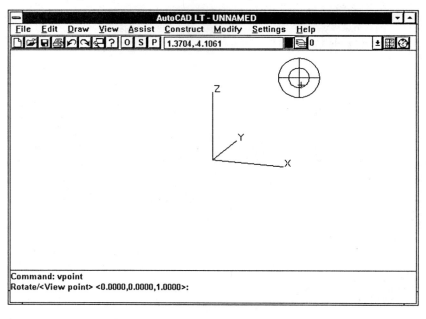

COMMAND OPTIONS
[Enter]   Bring up visual guides *(see figure)*.
Rotate   Indicate the new 3D viewpoint by angle.
<View point>   Indicate the new 3D viewpoint by x,y,z-coordinates.

RELATED AUTOCAD COMMANDS
- DdVpoint   Adjust viewpoint via dialogue box.
- DView   Changes the viewpoint of 3D objects, plus allows perspective mode.

RELATED SYSTEM VARIABLES
- VpointX, VpointY, VpointZ
            X,y,z-coordinates of current 3D view.
- WorldView Determines whether Vpoint coordinates are in WCS or UCS.

VpLayer

Rel. 1

Controls the visibility of layers in viewports when TileMode is turned off (*short for ViewPort LAYER*).

Command	Alt+	Menu bar	Alias	GC Alias
vplayer	V,L	[View] [Viewport Layer Visibility]	vl	...

```
Command: vplayer
?/Freeze/Thaw/Reset/Newfrz/Vpvisdflt:
Select a viewport: [pick]
```

COMMAND OPTIONS

Freeze — Indicate the names of layers to freeze in this viewport.
Newfrz — Create new layers, which will be frozen in newly created viewports (*short for NEW FReeZe*).
Reset — Resets the state of layers based on the Vpvisdflt settings.
Thaw — Indicate the names of layers to thaw in this viewport.
Vpvisdflt — Determines which layers will be frozen in a newly created viewport (*short for ViewPort VISibility DeFauLT*).
? — Lists the layers frozen in the current viewport.

RELATED AUTOCAD COMMANDS

- **DdLModes** Toggles the visibility of layers in viewports via a dialogue box.
- **Layer** Creates and controls layers in all viewports.
- **MView** Creates and joins viewports when tilemode is off.

RELATED SYSTEM VARIABLE

- **TileMode** Controls whether viewports are tiled or overlapping.

'View

Rel. 1

Saves and displays the view in the current viewport by name.

Command	Alt+	Menu bar	Alias	GC Aliases
'view	V,V	[View]	v	nv
		[View]		nx

```
Command: view
?/Delete/Restore/Save/Window: s
View name to save:
```

COMMAND OPTIONS

Delete	Delete a named view.
Restore	Restore a named view.
Save	Save the current view with a name.
Window	Save a windowed view with a name.
?	List the names of views saved in the current drawing.

RELATED AUTOCAD COMMANDS
- **DdView**   Creates and displays views via a dialogue box.
- **Rename**   Changes the names of views.

RELATED SYSTEM VARIABLES
- **ViewCtr**   The coordinates of the center of the view.
- **ViewSize**   The height of the view.

TIPS

- Name views in your drawing to quickly move from one detail to another.

- View names are up to 31 characters long and may not contain spaces.

- The **Plot** command plots named views of a drawing.

- Entities outside of the window created by the Window option may be displayed but are not plotted.

- You create separate views in model and paper space; when listing named views (with ?), AutoCAD indicates an "M" or "P" next to the view name.

Update *cont'd*

7. With the AutoCAD object in Write, you can do five things with the object, as listed below:
   - **Edit | Edit AutoCAD LT Drawing Object**
     Return to LT to edit the object.

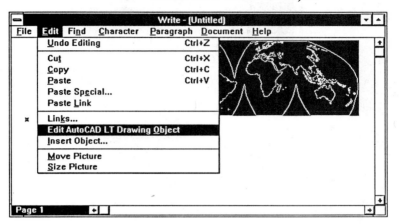

- **Edit | Move Picture**   Move the picture horizontally.
- **Edit | Size Picture**   Increase and decrease the size of picture.
- **Paragraph**   Position the picture to the left, center, or right.
- **[Del]**   Erase the object by pressing the **[Del]** key.

2. When the **Insert Object** dialogue box appears, select AutoCAD LT Drawing.

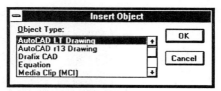

3. Click **OK**.
4. OLE launches AutoCAD LT automatically. Open a drawing or import a file.
5. From the menu bar, select **File | Update (Filename)**.

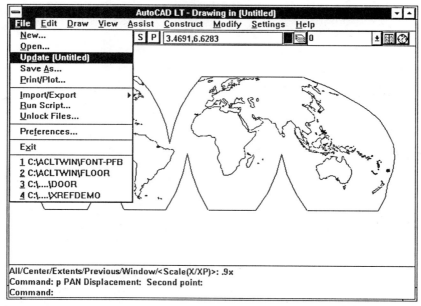

6. AutoCAD selects all objects in the drawing, sends them via OLE back to Write, and quits AutoCAD. The object appears in Write.

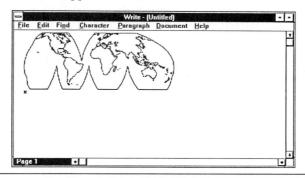

Update

Rel. 1

Updates linked or embeded OLE objects.

Command	Alt+	Menu bar	Alias	GC Alias
...	F,D	[File] [Update (Filename)]	...	...

Command can only be used at the menu bar.

COMMAND OPTIONS
None.

RELATED AUTOCAD COMMAND
- **CopyLink** Copies drawing to Windows Clipboard for linking in other applications.

RELATED WINDOWS COMMANDS
- **Edit | Insert Object** Places AutoCAD LT object (drawing) into application.
- **Edit | Edit Object** Returns object to AutoCAD LT for editing.

TIP
- OLE (object linking and embedding) is best avoided due to its instability and resource hogging.

QUICK START: Inserting an AutoCAD LT Object
How to place an AutoCAD LT drawing into another Windows application via OLE:

1. In the other application (such as Windows Write), select the **Edit | Insert Object** command from the menu bar.

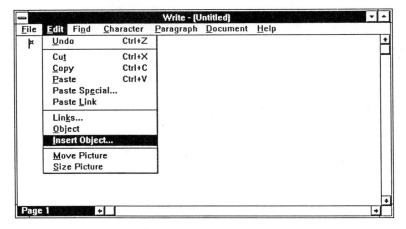

'Unlock

Rel. 1

Displays dialogue box to unlock locked (*.??K) files.

Command	Alt+	Menu bar	Alias	GC Aliases
'unlock	F,U	[File] [Unlock Files]	ul	...

Command: **unlock**
Displays dialogue box:

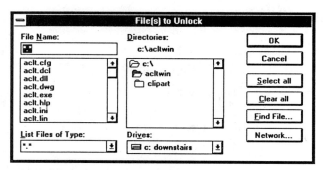

COMMAND OPTIONS

File Name	Replace *.* with *.??K to find locked files.
Select all	Select all files in subdirectory.
Clear all	Unselect all files.
Find File	Display Find File dialogue box.
Network	Display Connect Network dialogue box.

RELATED AUTOCAD COMMAND

- **Preferences** Toggle whether files are locked (default = on).

RELATED SYSTEM VARIABLES

None.

TIPS

- File locking ensures that no one on a network can use the files your copy of AutoCAD LT is currently employing.

- When using AutoCAD LT in a single user environment, turn off file locking.

Units *cont'd*

TIPS

- Because **Units** is a transparent command, you can use it to change units during another command.

- The 'Direction Angle' prompt lets AutoCAD start angle measurement from any direction.

- AutoCAD accepts the following notation for angle input:
    - <   Specify an angle based on current units setting.
    - <<   Bypass angle translation set by **Units** command to use 0-angle-is-east direction and decimal degrees.
    - <<< Bypass angle translation, use angle units set by **Units** command, and 0-angle-is-east direction.

- The system variable **UnitMode** forces AutoCAD to display units in the same manner that you enter them.

- Do not use a suffix (such as 'r' or 'g') for angles entered as radians or grads; instead, use the **Units** command to set angle measurement to radians and grads.

'Units

Rel. 1

Controls the display and format of coordinates and angles.

Command	Alt+	Menu bar	Alias	GC Aliases
'units	...	...	ut	nf
				un

Command: **units**

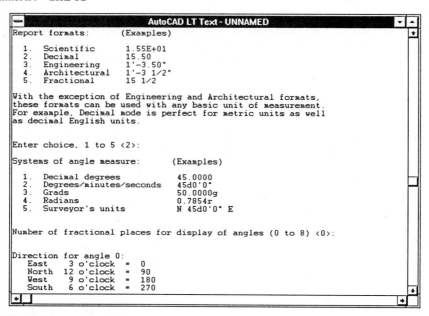

COMMAND OPTIONS
[F2]   Return to graphics window.

RELATED AUTOCAD COMMANDS
- **DdUnits**   Sets units via a dialogue box.
- **New**   Sets up a new drawing with appropriately scaled units.

RELATED SYSTEM VARIABLES
- **AngBase**   Direction of zero degrees.
- **AngDir**   Direction of angle measurement.
- **AUnits**   Units of angles.
- **AuPrec**   Displayed precision of angles.
- **LUnits**   Units of measurement.
- **LuPrec**   Displayed precision of coordinates.
- **UnitMode** Toggles type of display of units.

Undo

Rel. 1

Undo the effect of one or more previous commands.

Command	Alt+	Menu bar	Alias	GC Alias
undo	...	...	un	oo

Command: **undo**
Auto/Back/Control/End/Group/Mark/<number>:

COMMAND OPTIONS
Auto	Treats a menu macro as a single command.
Back	Undo back to the marker.
Control	Limits the options of the **Undo** command.
All	Toggles on full undo.
None	Turns off undo feature.
One	Limits the Undo command to a single undo.
End	Ends the group option.
Group	Group a sequence of operations.
Mark	Sets a marker.
<number>	Indicate the number of commands to undo.

RELATED AUTOCAD COMMANDS
- **Oops**   Unerases the most-recently erased object.
- **Quit**   Leaves the drawing without saving changes.
- **Redo**   Undoes the most recent undo.
- **U**   Single-step undo.

RELATED SYSTEM VARIABLES
- **UndoCtl**   Indicates the state of Undo:
  - 0 Undo is disabled.
  - 1 Undo is enabled.
  - 2 Single-command undo.
  - 3 Auto-group mode enabled.
  - 4 Group is currently active.
- **UndoMarks** Number of undo marks placed in the **Undo** control stream.

TIP
- Undo has no effect on the following commands: About, Area, AttExt, Delay, Dist, DxfOut, End, GraphScr, Help, Hide, Id, List, MSlide, New, Open, Plot, PsOut, QSave, Quit, Redraw, Regen, Resume, Save, SaveAs, Shade, and TextScr.

- The following commands have an **Undo** option built in: **Dim, Extend, Line, PEdit,** and **Trim**. These allow you to undo part of the effect of the command; using the **Undo** command after one of these undoes everything the command did.

UcsIcon

Rel. 1

Controls the location and display of the UCS icon.

Command	Alt+	Menu bar	Alias	GC Alias
ucsicon	A,C	[Assist] [UCS Icon]	ui	...

Command: **ucsicon**
ON/OFF/All/Noorigin/ORigin <ON>:

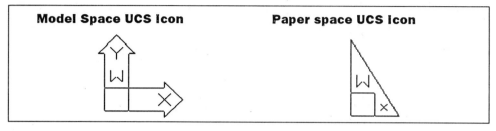

COMMAND OPTIONS
All   Makes **UcsIcon** command's changes apply to all viewports.
Noorigin   Always display UCS icon in lower-left corner.
OFF   Turn off display of UCS icon.
ON   Turn on display of UCS icon.
ORigin   Display UCS icon at the current UCS origin.

RELATED AUTOCAD COMMAND
- UCS   Creates and controls user-defined coordinate systems.

RELATED SYSTEM VARIABLE
- UcsIcon   Determines the display and origin of the UCS icon.

UCS *cont'd*

TIPS

- Use the **UCS** command to draw entities at odd angles in 3D space.

- Although you can create a UCS in paper space, you cannot use 3D viewing commands.

- A UCS can be aligned with these objects:
    - Point, line, trace, 2D polyline, and solid.
    - Arc and circle.
    - Text, shape, dimension, and attribute definition.
    - Block reference.

DEFINITIONS

UCS:
- User-defined 2D coordinate system oriented in 3D space.
- Sets a working plane, orients 2D objects, defines the extrusion direction, and the axis of rotation.

WCS:
- World coordinate system.
- The default 3D x,y,z-coordinate system.

 # Ucs

Rel. 1

Defines a new coordinate plane (*short for User Coordinate System*).

Command	Alt+	Menu bar	Alias	GC Aliases
UCS	A, U	[Assist]	...	do
		[Set UCS]		mr

```
Command: ucs
Origin/ZAxis/3point/Entity/View/X/Y/Z/Prev/Restore/Save/Del/
    ?/<World>:
```

COMMAND OPTIONS
Del	Delete the name of a saved UCS.
Entity	Align UCS with a picked entity.
Origin	Move the UCS to a new origin point.
Prev	Restore the previous UCS orientation.
Restore	Restore a named UCS.
Save	Save the current UCS by name.
View	Align the UCS with the current view.
<World>	Align the UCS with the WCS.
X	Rotate the UCS about the x-axis.
Y	Rotate the UCS about the y-axis.
Z	Rotate the UCS about the z-axis.
ZAxis	Align the UCS with a new origin and z-axis.
3point	Align the UCS with a point on the positive x-axis and positive x,y-plane.
?	List the names of saved UCS orientations.

RELATED AUTOCAD COMMANDS
- **DdUcs**   Modifies the UCS via a dialogue box.
- **UcsIcon**   Controls the visibility of the UCS icon.
- **Plan**   Changes the view to the plan view of the current UCS.

RELATED SYSTEM VARIABLES
- **UcsFollow**   Automatically shows plan view in new UCS.
- **UcsIcon**   Determines visibility and location of UCS icon.
- **UcsOrg**   WCS coordinates of UCS icon.
- **UcsXdir**   X-direction of current UCS.
- **UcsYdir**   Y-direction of current UCS.
- **WorldUcs**   Correlation of WCS and UCS.

U

Rel. 1

Undoes the most recent AutoCAD command (*short for Undo*).

Command	Alt+	Menu bar	Alias	GC Alias
u	E,U	[Edit] [Undo]	...	...

Command: **u**

COMMAND OPTIONS
None.

RELATED AUTOCAD COMMANDS
- **Oops** — Unerases the most recently erased object.
- **Quit** — Exits the drawing, undoing all changes.
- **Redo** — Redoes the most-recent undo.
- **Undo** — Allows more sophisticated control over undo.

RELATED SYSTEM VARIABLE
- **UndoCtl** — Determines the state of undo control.

TIPS
- The **U** command is convenient for stepping back through the design process, undoing one command at a time.

- The **U** command is the same as the **Undo 1** command; thus, for greater control over the undo process, use the **Undo** command.

- The **Redo** command redoes the most-recent undo only.

- The **Quit** command restores the drawing to its original state.

- The undo mechanism creates a mirror drawing file on disk; when your computer is low on disk space, however, the **Undo** command cannot be disabled with system variable **UndoCtl** (which is read-only) as in full-size AutoCAD.

- Commands that involve writing to file, plotting, and system variables are not undone.

Trim

Rel. 2

Trims lines, arcs, circles, and 2D polylines back to a real or projected cutting line or view.

Command	Alt+	Menu bar	Alias	GC Aliases
trim	M,T	[Modify] [Trim]	tr	it mt rn

```
Command: trim
Select cutting edges...
Select objects: [pick]
Select objects: [Enter]
<Select object to trim>/Undo: [pick]
```

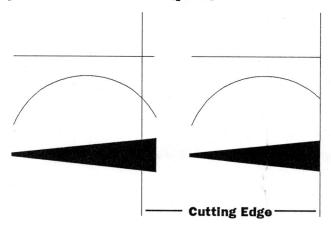

——— Cutting Edge ———

COMMAND OPTIONS
Select object to trim
    Pick the objects at the trim end.
Undo   Untrim the last trim action.

RELATED AUTOCAD COMMANDS
- **Change**   Changes the size of lines, arcs, and circles.
- **Extend**   Lengthens lines, arcs, and polylines.
- **PEdit**   Changes polylines.
- **Stretch**   Lengthens or shortens lines, arcs, and polylines.

TIP
- To trim a pair of intersecting lines, it can be faster to use the **Fillet** command with radius = 0.

cons supplied with AutoCAD LT Release 2:

Arc		Extend		Save	
Area		Fillet		Scale	
Array		Hatch		Stretch	
BHatch		HatchEdit		Text	
Block		Help		Toolbox	
Boundary		Id		Trim	
Break		Insert		Ucs	
Chamfer		Layer		Undo	
ChProp		Line		Zoom	
Circle		List			
Copy		Measure		Center	
CopyClip		Mirror		Endpoint	
CopyLink		Move		Insertion	
DdEdit		New		Intersection	
DdEmodes		Offset		Midpoint	
DDim		Open		Node	
DdModify		OSnap		None	
DdRModes		Pan		Perpendicular	
DdUcs		PasteClip		Quadrant	
DdUcsP		PEdit		Tangent	
Dim		PLine			
Dist		Plot		Tracking	
Divide		Point		Close	
DLine		Polygon		[Ctrl]+C	
Donut		Preferences		@	
DsViewer		Rectang		.X	
Ellipse		Redo		.Y	
Erase		Redraw		.Z	
Explode		Rotate			

Image Name	Select the name of an icon (displayed in center).
Command	Type the macro for this button.
Save	Save changes to initialization file Aclt.Ini.
Toolbox Width	Determine width of toolbox; height is calculated automatically:
Floating	Number of buttons wide when floating (default = 12).
Locked	Number of buttons wide when fixed (default = 4).
OK	Save changes and dismiss dialogue box.
Insert	Insert as a new button.
Delete	Remove button from toolbox.
Cancel	Cancel changes and dismiss dialogue box.
Help	Display help window.

RELATED AUTOCAD COMMAND
- **Preferences** Determines whether toolbox appears when AutoCAD LT starts.

RELATED FILE
- **Aclt.Ini** The initialization file contains toolbox macros.

TIPS
- To display a tooltip, place cursor over toolbox button.

- To edit all macros attached to buttons, edit the **[AutoCAD LT ToolBox]** section of the Aclt.Ini file with a text editor, such as Notepad:

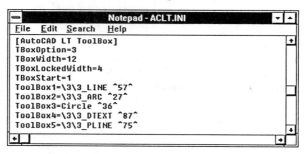

Toolbox

Rel. 1

Displays the toolbox in three modes: floating, fixed, and off.

Command	Alt+	Menu bar	Alias	GC Alias
toolbox	S,B	[Settings]	tl	...
		[Toolbox Style]		

Command: `toolbox`

*The toolbox changes position each time the **Toolbox** command is used: 1. Floating; 2. Fixed to left side; 3. Turned off; 4. Fixed to right side; and back to floating, as shown below:*

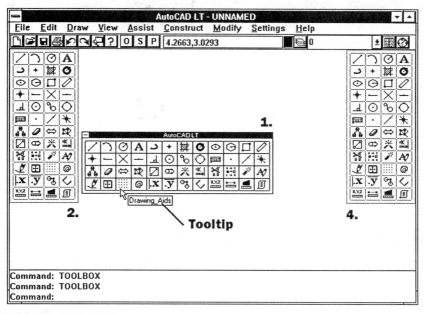

COMMAND OPTION

🖱 **[Right-click]** Change icon and macro via the **Toolbox Customization** dialogue box:

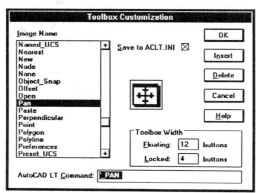

'Time

Rel. 1

Display timely information about the current drawing.

Command	Alt+	Menu bar	Alias	GC Alias
'time	A,T	[Assist] [Time]	ti	...

Command: **time**
Sample output:

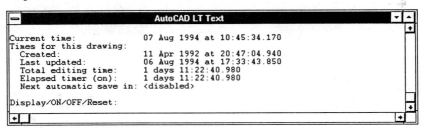

COMMAND OPTIONS

Display	Display the current time information.
OFF	Turn the user timer off.
ON	Turn the user timer on.
Reset	Reset the user timer.
[F2]	Return to graphics screen.

RELATED AUTOCAD COMMAND
- Status   Displays information about the current drawing and environment.

RELATED SYSTEM VARIABLES
- CDate   The current date and time.
- Date   The current date and time in Julian format.
- TdCreate   Date and time the drawing was created.
- TdInDwg   The time the drawing spent in AutoCAD.
- TdUpdate   The last date and time the drawing was changed.
- TdUsrTimer The current user timer setting.
- SaveTime   The automatic drawing save interval.

TIPS
- The times displayed by the **Time** command are only accurate when the computer's clock is accurate; unfortunately, the clock in most personal computers strays by many minutes per week.
- Set the automatic save time with the **Preferences** command.

'TextScr

Rel. 1

Switches from the graphics window to the text window (*short for TEXT SCReen*).

Command	Alt+	Menu bar	Alias	GC Alias
'textscr	E,X	[Edit] [Text Window]	[F2]	...

Command: **textscr**

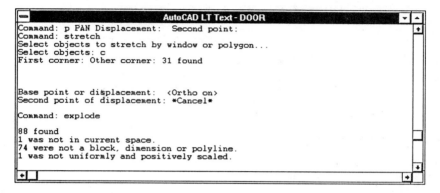

COMMAND OPTIONS
None.

RELATED AUTOCAD COMMAND
■ **GraphScr**   Switches from text window to graphics window.

RELATED SYSTEM VARIABLE
■ **ScreenMode** Reports whether screen is in text or graphics mode.
    0 Text window.
    1 Graphics window.
    2 Dual screen displaying both text and graphics.

TIP
■ **TextScr** does not work with dual-screen systems.

?   List the currently loaded styles:

```
─                     AutoCAD LT Text - UNNAMED                     ▼ ▲
Text style(s) to list <*>:
Text styles:

Style name: STANDARD       Font files: txt
   Height: 0.0000   Width factor: 1.0000   Obliquing angle: 0
   Generation: Normal

Current text style: STANDARD
Justify/Style/<Start point>:
```

RELATED AUTOCAD COMMANDS
- **Change**   Changes the text height, rotation, style, and content.
- **DdModify** Changes all aspects of the style applied to a line of text.
- **DText**   Places new text in the drawing interactively.
- **Style**   Creates new text styles.

RELATED SYSTEM VARIABLES
- **TextSize** The current height of text.
- **TextStyle** The current style.

TIPS
- The **Text** command does not prompt for the text height when:
    - The style has a text height other than 0.0.
    - Fit justification is selected.

- Use the **QText** command to turn lines of text into rectangles, which redraw much faster than text; turning on **QText** does affect text during plotting; qtext blocks are plotted as text.

 Text

Rel. 1

Places one line of text in the drawing.

Command	Alt+	Menu bar	Alias	GC Alias
text	...	...	tx	tp

```
Command: text
Justify/Style/<Start point>: j
Align/Fit/Center/Middle/Right: r
Height <0.2000>: [Enter]
Rotation angle <0>: [Enter]
Text:
```

COMMAND OPTIONS

[Enter] Contines text one line below previously placed text line.
Justify Displays the justification options (default = left):

Text Insertion point Left
 Aligned
  Fitted
  Center
   Middle
  Right

Align   Align the text between two points; text height is maintained.
Fit   Fit the text between two points; text height is adjusted.
Center   Center the text along the baseline.
Middle   Center the text horizontally and vertically.
Right   Right-justify the text.

Start point Indicate the text insertion point.
Style   Displays the style submenu:
  Style name   Indicate a different style name.

PostScript fonts provided with AutoCAD LT Release 2 (stored in PFB and PFM files):

cibt____.pfb	*Cibt____:*	ABC abc 123 - + & $
cobt____.pfb	*Cobt____:*	ABC abc 123 - + & #
euro____.pfb	Eur____:	ABC abc 123 - + & $
eur____.pfb	*Euro____:*	ABC abc 123 - + & $
par____.pfb	Par	ABC abc 123 ~ & $
romb____.pfb	Rom____:	ABC abc 123 - + & $
romi____.pfb	Romb____:	ABC abc 123 - + & $
rom____.pfb	*Romi____:*	ABC abc 123 - + & $
sasbo____.pfb	Sas____:	ABC abc 123 - + & $
sasb____.pfb	Sasb____:	ABC abc 123 - + & $
saso____.pfb	*Sasbo____:*	ABC abc 123 - + & $
sas____.pfb	*Saso____:*	ABC abc 123 - + & $
suf____.pfb	Suf	ABC abc 123 - ¿ & $
teb____.pfb	Te____:	ABC ABC 123 - + & $
tel____.pfb	Teb____:	ABC ABC 123 - + & $
te____.pfb	Tel____:	ABC ABC 123 - + & $

Style *cont'd*

Vector fonts provided with AutoCAD LT Release 2 (stored in SHX files):

complex.shx	Complex: ABC abc 123 — + & $
cyrillic.shx	Вшсиллив: АБВ абв 123 — + & $
cyriltlc.shx	Чйрилтлч: АБЧ абч 123 — + & $
gothice.shx	GothicE ABC abc 123 — + & $
gothicg.shx	GothicG: ABC abc 123 — + & $
gothici.shx	GothicI: ABC abc 123 — + & $
greekc.shx	ГрεεκX: ABX αβχ 123 — + & $
greeks.shx	ГрεεκΣ: ABX αβχ 123 — + & $
iso9.shx	Iso9: ABC abc 123 - + & $
italic.shx	*Italic: ABC abc 123 — + & $*
italicc.shx	*ItalicC: ABC abc 123 — + & $*
italict.shx	*ItalicT: ABC abc 123 — + & $*
monotxt.shx	MonoTxt: ABC abc 123 — + & $
romanc.shx	RomanC: ABC abc 123 — + & $
romand.shx	RomanD: ABC abc 123 — + & $
romans.shx	RomanS: ABC abc 123 — + & $
romant.shx	RomanT: ABC abc 123 — + & $
scriptc.shx	*ScriptC: ABC abc 123 — + & $*
scripts.shx	*ScriptS: ABC abc 123 — + & $*
simplex.shx	Simplex: ABC abc 123 — + & $
syastro.shx	♋→∂´∪ℒ: ⊙♀♂ ✳'' 123 — + & $
symap.shx	(map symbols) 123 — + & $
symath.shx	√§∏∃√∫: ℵ' ←↓∂ 123 — + & $
symeteo.shx	(meteo symbols) 123 — + & $
symusic.shx	(music symbols) 123 — + & $
txt.shx	Txt: ABC abc 123 — + & $

Obliquing angle Angle of individual characters in line of text (default = 0).
Backward   Print text forwards (default) or backwards.

    Backwards text
    Upside-down text *(shown inverted)*

Upside-down   Print text rightside-up (default) or upside-down.
Vertical   Print text vertically (characters maintain horizontal orientation).

RELATED AUTOCAD COMMANDS
- **Change**   Changes the style assigned to selected text.
- **DText**   Places text using current style; allows change of style.
- **Purge**   Removes unused text style definitions.
- **Rename**   Renames a text style name.
- **Text**   Enters text using current style; allows change of style.

RELATED SYSTEM VARIABLES
- **TextStyle** The current text style.
- **TextSize**   The current text height.

RELATED FILES
- ***.SHX**   Autodesk's format for vector fonts.
- ***.PFB**   PostScript font definition files.
- ***.PFM**   PostScript font metrication files.

TIPS

- Not all fonts support the **Vertical** option; for example, none of the PostScript fonts.

- Use a **Width factor** of 0.85 to fit 15% more text in the same space without affecting readability.

- To make plain fonts better looking, set the **Obliquing angle** to +15 degrees.

- The fastest displaying text font is TXT; the slowest text fonts are PostScript.

- The best font is Simplex, which combines legibility with speedy display.

- If you have difficulties with PostScript fonts plotting incorrectly, comment out the font names in file Aclt.Psf by prefixing the font names with a ; (semi-colon).

'Style

Rel. 1

Creates and modifies text styles based on a font file.

Command	Alias	Menu bar	Alias	GC Alias
'style	S,T	[Settings] [Text Style]	st	...

Command: **style**
Select font file from dialogue box:

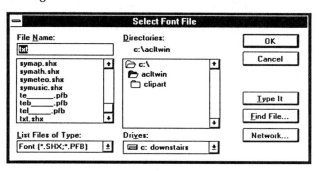

```
Text style name (or ?) <STANDARD>:
Font file <ROMANS>:
 Height <0.0000>: [Enter]
 Width factor <1.00>: [Enter]
Obliquing angle <0>: [Enter]
Backwards? <N> [Enter]
Upside-down? <N> [Enter]
Vertical? <N> [Enter]
STANDARD is now the current text style.
```

COMMAND OPTIONS

Text style name   Name of the style (default = 'STANDARD').
?   List names of style defined in current drawing.
Height   Height of text from baseline to top of UPPERCASE character.
Width factor = 0.5
Width factor = 1.0
Width factor = 1.5

Width factor   Relative width of characters (default = 1.0).
Obliquing angle = −45
Obliquing angle = 0
Obliquing angle = 45

TIPS

■ The effect of the **Stretch** command is not always obvious; be prepared to use the **Undo** command.

■ The first time you select objects for the **Stretch** command, you must use **Crossing** object selection — despite what the prompt tells you.

■ Objects entirely within the selection window are moved, rather than stretched.

■ The **Stretch** command will not move a hatch pattern unless the hatch's origin is included in the selection set.

■ Use the **Stretch** command to automatically update associative dimensions by including the dimension's endpoint in the selection set.

 # Stretch

Rel. 1

Stretches objects to lengthen, shorten, or distort them.

Command	Alias	Menu bar	Alias	GC Alias
stretch	M,S	[Modify]	s	ss
		[Stretch]		ws

```
Command: stretch
Select objects to stretch by window of polygon...
Select objects: c
First corner: [pick]
Other corner: [pick]
Select objects: [Enter]
Base point: [pick]
New point: [pick]
```

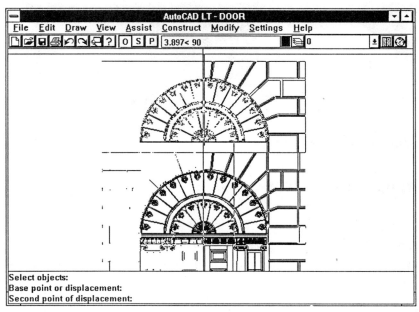

COMMAND OPTIONS
None.

RELATED AUTOCAD COMMANDS
- **Change** Changes the size of lines, circles, text, blocks, and arcs.
- **Scale**   Increases or decreases the size of any object.

Solid

Rel. 1

Draws solid filled triangles and quadrilaterals; does *not* create a 3D solid model.

Command	Alt+	Menu bar	Alias	GC Alias
solid	D,S	[Draw] [Solid]	so	...

Command: **solid**
First point: **[pick]**
Second point: **[pick]**
Third point: **[pick]**
Fourth point: **[pick]**
Third point: **[Enter]**

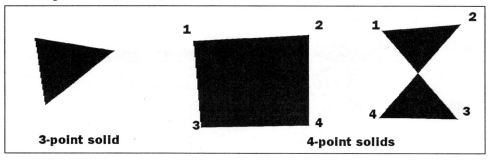

3-point solid   4-point solids

COMMAND OPTIONS
None.

RELATED AUTOCAD COMMANDS
- Fill   Turns object fill off and on.
- PLine   Draws polylines and polyline arcs with width.

RELATED SYSTEM VARIABLE
- FillMode Determines whether solids are displayed filled or outlined.

TIPS
- To draw a four-sided solid requires care with the picking order: see figure above.

- Since large solid areas are slow to redraw, the the **Fill** command to display just the solid outlines.

[S] 'Snap   Rel. 1

Sets the drawing resolution, grid origin, isometric mode, and angle.

Command	Ctrl+	Function Key	Alias	GC Aliases
'snap	B	[F9]	sn	cm
				oa
				sg

Command: **snap**
Snap spacing or ON/OFF/Aspect/Rotate/Style <1.0000>:

COMMAND OPTIONS
Snap spacing Set the snap increment.
OFF Turn snap off.
ON Turn snap on.
Aspect Set separate x and y snap increments:
  Horizontal spacing Snap spacing in x-direction (default = 1).
  Vertical spacing Snap spacing in y-direction (default = 1).

Rotate Rotate the cross-hairs for snap and grid.
  Base point Specify origin for rotated snap (default = 0,0).
  Rotation angle Angle of rotation (default = 0 degrees).

Style Switch between standard and isometric style: 'Standard/Isometric:'.

RELATED AUTOCAD COMMANDS
- **DdRModes** Set snap values via a dialogue box.
- **Grid** Turn on the grid.
- **Isoplane** Switch to a different isometric drawing plane.

RELATED KEYSTROKES
- **[Ctrl]+B** Toggle snap mode between on and off.
- **[F9]** Toggle snap mode between on and off.

RELATED SYSTEM VARIABLES
- **SnapAng** Current angle of the snap rotation.
- **SnapBase** Base point of the snap rotation.
- **SnapIsopair** Current isometric plane setting.
- **SnapMode** Determines whether snap is on.
- **SnapStyl** Determines style of snap.
- **SnapUnit** The current snap increment in x- and y-directions.

TIPS
- Turn on the grid (with function key **[F7]**) to see the snap points.
- Set the snap distance to the smallest drawing distance.

RELATED SYSTEM VARIABLES
- **ShadeDif**   Determines the shading contrast (default=70).
- **ShadEdge**   Determines the style of shading:
    - 0 256-color shading.
    - 1 256-color shading with outlined polygons.
    - 2 Hidden-line removal.
    - 3 16-color shading (default).

ShadEdge = 0

ShadEdge = 1

ShadEdge = 2

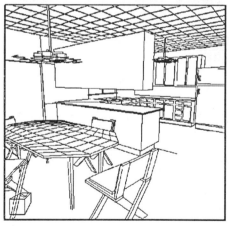

ShadEdge = 3

TIPS
- The smaller the viewport, the faster the rendering.
- The shading is controlled by system variables **ShadEdge** and **ShadeDif**.

Shade

Rel. 1

Performs 16- and 256-color shaded renderings, and quick hidden-line removal of 3D drawings.

Command	Alt+	Menu bar	Alias	GC Alias
shade	V,S	[View] [Shade]	sh	...

```
Command: shade
Regenerating drawing.
Shading in 9 passes.
Shading complete.
```

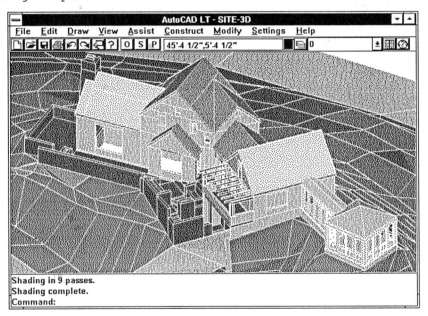

COMMAND OPTIONS
None.

RELATED AUTOCAD COMMANDS
- **CopyImage** Copies the rendering to the Clipboard in raster format.
- **DView** Does hidden-line removal of perspective views.
- **Hide** Does true hidden-line removal of 3D drawings.
- **MSlide** Saves a rendered view as an SLD-format slide file.
- **MView** Does a hidden-line view of individual viewports during plots and prints.
- **Plot** Does a hidden-line view during plotting.

'SetVar

Rel. 1

Lists the settings of system variables; allows you to change variables that are not read-only (*short for SET VARiable*).

Command	Alt+	Menu bar	Alias	GC Alias
'setvar	...	...	...	...

```
Command: setvar
Variable name or ?:
```

Example usage:
```
Command: setvar
Variable name or ?: highlight
New value for HIGHLIGHT <0>: 1
```

COMMAND OPTIONS
Variable name Indicate the system variable name you want to access.
? Lists the names and settings of system variables.

RELATED AUTCAD COMMANDS
- **GetEnv**   Displays the values of variables in the Aclt.Ini initialization file.
- **SetEnv**   Changes the values of variables in the Aclt.Ini initialization file.

TIPS
- See Appendix A for the complete list of all system variables found in AutoCAD LT Release 2.

- Almost all system variables can be entered without the **SetVar** command. For example,
  ```
  Command: highlight
  New value for HIGHLIGHT <0>: 1
  ```

- The following system variables are not listed by the **SetVar** command. These system variables are used by third-party programmers, for debugging, or are obsolete:
  - _PkSer and _Server.
  - AxisMode and AxisUnit.
  - EntExts and ErrNo.
  - Flatland.
  - MacroTrace.
  - PHandle.
  - QaFlags.
  - UserI1 through UserI5 and UserR1 through UserR5.

'SetEnv

Rel. 1

Changes the value of variables in part of the Aclt.Ini initialization file.

Command	Alt+	Menu bar	Alias'	GC Alias
'setenv	...	...	...	...

```
Command: setenv
Variable name:
```

Example output:
```
Command: setenv
Variable name: monovectors
Value <0>: 1
```

COMMAND OPTIONS
All variables in [AutoCAD LT General] section of Aclt.Ini file.

RELATED AUTOCAD COMMANDS
- **GetEnv** Displays the values of variables in the Aclt.Ini file.
- **Preferences** Dialogue box interface for setting some INI variables.
- **SetVar** Reads and changes the values of system variables.

RELATED AUTOCAD FILES
- **Aclt.Ini** Contains the settings for AutoCAD's Preferences and toolbox macros.

TIPS
- The **SetEnv** command only works with the [AutoCAD LT General] section of the Aclt.Ini file.

- If the variable does not exist, **SetEnv** creates it, as follows:
  ```
  Command: setenv
  Variable name: asdf
  Value: 1234
  ```
  The new variable is appended to the end of the [AutoCAD LT General] section, as follows:
  ```
  asdf=1234
  ```

- You cannot have spaces in the value, which AutoCAD interprets as [Enter].

- See Appendix B for the variables in the Aclt.Ini file.

- The **SetEnv** command does not exist in AutoCAD Release 12 or 13.

Select

Rel. 1

Creates a selection set of objects before executing a command.

Command	Alt+	Menu bar	Alias	GC Aliases
select	A,S	[Assist]	se	se
		[Select]		fe

Command: **select**
Select objects: **[pick]**

COMMAND OPTIONS

A	Continue to add entities after using the R option (*short for Add*).
AU	Switches from [pick] to C or W modes, depending on whether an entity is found at the initial pick point (*short for AUtomatic*).
ALL	Select all objects in the drawing.
BOX	Goes into C or W mode, depending on how the cursor moves.
C	Selects objects in and crossing the selection box (*Crossing*).
CP	Selects all objects inside and crossing the selection polygon.
F	Selects all objects crossing a polyline (*short for Fence*).
L	Selects the last-drawn object still visible on the screen (*Last*).
M	Make multiple selections before AutoCAD scans the drawing; saves time in a large drawing (*short for Multiple*).
P	Selects the previously selected objects (*short for Previous*).
R	Remove entities from the selection set (*short for Remove*).
SI	Select only a single set of objects before terminating **Select** command (*short for SIngle*).
U	Removes the most-recently added selected objects (*short for Undo*).
W	Selects all objects inside the selection box (*short for Window*).
WP	Selects all objects inside the selection polygon.
[pick]	Select a single object.
[Enter]	Exit the **Select** command.
[Ctrl]+C	Stop the **Select** command.

RELATED AUTOCAD COMMANDS

All commands that prompt 'Select objects:'.

RELATED SYSTEM VARIABLES

- **PickAdd**   Controls how entities are added to a selection set.
- **PickAuto**   Controls automatic windowing at the 'Select Objects:' prompt.
- **PickDrag**   Controls method of creating a selection box.
- **PickFirst**   Controls command-object selection order.

Script *cont'd*

QUICK START: Writing a script file.

1. A script file consists of the exact keystrokes you type for a command. The script file must be plain ASCII text, so write the script file using a text editor, rather than a word processor.

2. Here is an example script that places a door symbol in the drawing:
   ```
   ; Inserts DOOR2436 symbol at x,y = (76,100)
   ; x-scale = 0.5, y-scale = 1.0, rotation = 90 degrees
   insert door2436 76,100 0.5 1.0 90
   ```

3. In the script, these characters have special meaning:
   - ■ (Space or end-of-line) Equivalent to pressing the spacebar or [Enter] key.
   - ■ ; (Semi-colon) Include a comment in the script file.
   - ■ * (Asterisk) Prefix the **VSlide** command to preload it SLD file.

4. Save script file with any 8-character file name and the .SCR extension. For this example, use 'InsDoor.Scr'.

5. Return to AutoCAD and run the script with the **Script** command:
   ```
   Command: script
   Script file: insdoor.scr
   Command: insert
   Block name (or ?): door2436
   Insertion point: 76,100
   X scale factor <1>/Corner/XYZ: 0.5
   Y scale factor (default = X):1.0
   Rotation angle <0>:90
   ```

6. Rerun the script with the **RScript** command.

'Script

Rel. 1

Runs an ASCII file containing a sequence of AutoCAD instructions to automatically execute a series of commands.

Command	Alt+	Menu bar	Alias	GC Alias
'script	F,R	[File] [Run Script]	sr	...

Command: **script**
Displays Select Script dialogue box:

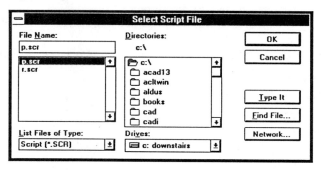

COMMAND OPTIONS
[Backspace]   Interrupts the script.
[Ctrl]+C   Stops the script.

RELATED AUTOCAD COMMANDS
- **Delay**   Pauses, in milliseconds, before executing the next command.
- **Resume**   Resumes a script after a script has been interrupted.
- **RScript** Repeats a script file.

TIPS
- Since the **Script** command is a transparent command, it can be used during another command.

- Prefix the **VSlide** comand to preload it into memory; this results in a faster slide show:
     *vslide

- AutoCAD can start with a script file on the command line:
     C:\ **acad13 dwgname scrname**

Since the script filename must follow the drawing filename, use a dummy drawing filename, such as 'X'.

Scale

Rel. 1

Changes the size of selected entities, to make them smaller or larger.

Command	Alt+	Menu bar	Alias	GC Aliases
scale	M,C	[Modify]	sc	sz
		[Scale]		wz

```
Command: scale
Select objects: [pick]
Select objects: [Enter]
Base point: [pick]
<Scale factor>/Reference: r
Reference length <1>:
```

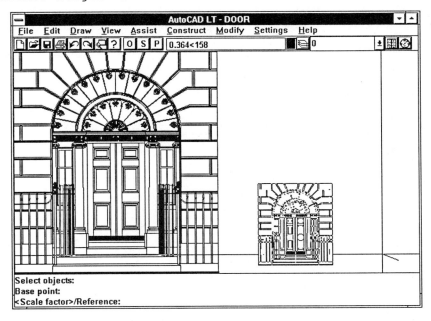

COMMAND OPTIONS

Reference   Supply a reference value.
<Scale factor>   Indicate scale factor, which applies equally in the x-, y- and z-directions.

RELATED AUTOCAD COMMANDS

- Insert   Allows a block to be scaled independently in the x-, y-, and z-directions.
- Plot   Allows a drawing to be plotted at any scale.

- The **SaveDIB** command is unusual because it only saves portions of the AutoCAD LT window, as follows:
    - Toolbar.
    - Entire drawing area.
    - Cursor.
    - 'Command:' prompt area.

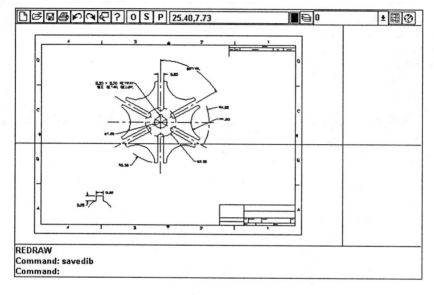

SaveDIB

Rel. 1

Saves part of the AutoCAD LT window and drawing to disk as a BMP-format raster file (*short for SAVE Device Independent Bitmap*).

Command	Alt+	Menu bar	Alias	GC Alias
savedib	F,I,B	[File] [Import/Export] [BMP Out]	...	...

Command: **savedib**
Displays dialogue box:

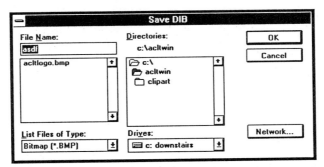

COMMAND OPTIONS
None.

RELATED AUTOCAD COMMANDS
- **CopyImage** Copies drawing to Windows Clipboard in BMP raster format.
- **Plot** Exports just the drawing in BMP and other raster formats.

RELATED WINDOWS COMMANDS
- **[Prt Scr]** Copy the entire screen to the Windows Clipboard.
- **[Alt]+[Prt Scr]** Copy the topmost window to the Clipboard.

TIPS
- For copying the entire AutoCAD LT window, the **[Alt]+[Prt Scr]** keystroke is a better alternative to the **SaveDIB** command.

- For exporting just the drawing in BMP format, the **CopyImage** command is a better alternative to the **SaveDIB** command.

- *Warning*: the **SaveDIB** command creates very large files; for example, with AutoCAD LT full screen at 1024x768 resolution, the BMP file is 2.2MB in size.

- Depsite its name, the **SaveDIB** command saves the image in BMP format, not DIB format.

SaveAs

Rel. 1

Saves the current drawing to disk as a Release 12-format DWG drawing file; when you save the drawing with different filename, the drawing takes on the new name.

Command	Alt+	Menu bar	Alias	GC Alias
saveas	F,A	[File] [Save As]	ss	dn

Command: **saveas**
Displays dialogue box:

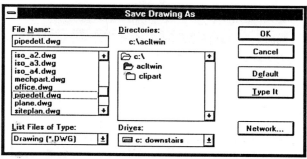

COMMAND OPTIONS
OK   Accept new filename
Cancel   Cancel **SaveAs** command and return to drawing.
Default   Use current filename.
Type It   Dismisses dialogue box and prompts:
             `Save current changes as <filename>:`
Network   Display **Connect Network Drive** dialogue box.

RELATED AUTOCAD COMMANDS
- End   Saves the drawing and exits AutoCAD.
- Quit   Exits AutoCAD without saving the drawing.
- Save   Saves the drawing with the current name.

RELATED SYSTEM VARIABLE
- DwgName   Name of the drawing; 'UNNAMED' when unnamed.

 # Save

Rel. 1

Saves the drawing to disk.

Command	Alt+	Menu bar	Alias	GC Alias
save	...	...	sa	ds

Command: **save**

COMMAND OPTIONS
None.

RELATED AUTOCAD COMMANDS
- End   Saves the drawing and exits AutoCAD.
- Quit   Exits AutoCAD without saving the drawing.
- QSave   Saves drawing without prompting for name.

RELATED SYSTEM VARIABLE
- DwgName Name of the drawing; 'UNNAMED' when unnamed.

TIP
- When the drawing is unnamed, the **Save** command mimics the **SaveAs** command and displays the **Save Drawing As** dialogue box.

'RScript

Rel. 1

Repeats the currently loaded script file (*short for Repeat SCRIPT*).

Command	Alt+	Menu bar	Alias	GC Alias
'rscript	...	...	...	...

Command: **rscript**

COMMAND OPTIONS
None.

RELATED AUTOCAD COMMANDS
- **Resume**   Resumes a script file after being interrupted.
- **Script**   Loads and runs a script file.

TIP
- Only one script at a time can be loaded in AutoCAD. Thus, the **RScript** command reruns whatever script is currently loaded.

# Rotate

Rel. 1

Rotates objects about a base point in a 2D plane.

Command	Alt+	Menu bar	Alias	GC Aliases
rotate	M,R	[Modify]	ro	dr
		[Rotate]		ro

```
Command: rotate
Select objects: [pick]
Select objects: [Enter]
Base point: [pick]
<Rotation angle>/Reference: R
Reference angle <0>:
New angle:
```

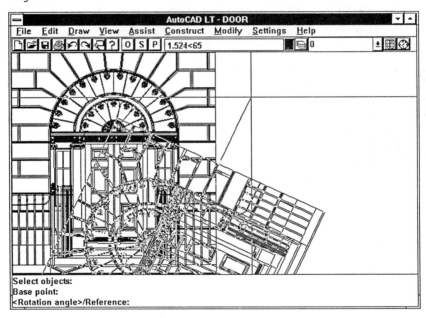

COMMAND OPTIONS
<Rotation angle> Specify the angle of rotation.
Reference Indicate a starting reference angle and an ending reference angle.

RELATED AUTOCAD COMMAND
- **Change** Rotates text entities.

RELATED AUTOCAD COMMAND
- New   Places drawing border and revdate information.

RELATED SYSTEM VARIABLE
- DwgName   Current drawing name (default = 'UNNAMED').

TIPS
- The first time you use the **RevDate** command in a drawing, it obtains the following information from:
    - Name   Initialization file, Aclt.Ini.
    - Date   The system date.
    - Time   The system clock.
    - Drawing   System variable DwgName.

- Subsequent times that you use the **RevDate** command, it updates the information.

- If the computer's date and clock are incorrect, then **RevDate** places the incorrect date and time in the drawing.

- The revdate info is placed as a block in the drawing:
    - **Block name**   'REVINFO'
    - **Layer name**   'TITLE-BLOCK'
    - **Attribute tags**:
        - USER   Holds the user name and company.
        - REVDATE Holds the date and time.
        - FNAME   Holds the drawing's filename.

RevDate

Rel. 2

Places or updates revision information in the drawing as a block: registered name, date, time, and drawing name (*short for REVision DATE*).

Command	Alt+	Menu bar	Alias	GC Alias
revdate	...	[Modify] [Date and Time]	...	...

First time command is used in drawing:
```
Command: revdate
REVDATE block insertion point: [pick]
REVDATE block rotation (0 or 90 degrees) <0>: [Enter]
```

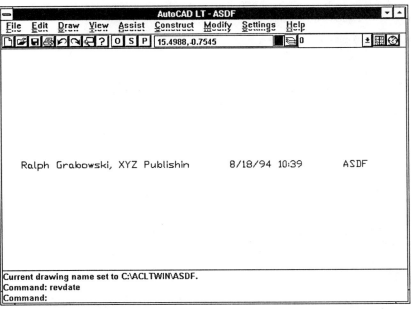

Subsequent times command is used:
```
Command: revdate
```
Date, time, and drawing name are updated.

COMMAND OPTIONS

Insertion point Indicate the point to place the revision info.
Rotation Specify rotation angle for revision info:
  0 Zero degrees: reads from left to right (default).
  90 Sideways: reads from bottom to top.

'Resume

Rel. 1

Resumes a script file after pausing it by pressing the **[Backspace]** key.

Command	Alt+	Menu bar	Alias	GC Alias
'resume	...	...	...	...

Command: **resume**

COMMAND OPTIONS
[Backspace]   Pause the script file.
[Ctrl]+C   Stop script file.

RELATED AUTOCAD COMMANDS
- RScript   Reruns the current script file.
- Script   Loads and runs a script file.

RELATED SYSTEM VARIABLES
- *None.*

Rename

Rel. 1

Allows you to change the names of blocks, dimension styles, layer, linetypes, text styles, UCS names, views, and viewports.

Command	Alt+	Menu bar	Alias	GC Alias
rename	...	...	rn	...

Command: **rename**
Block/Dimstyle/LAyer/LType/Style/Ucs/VIew/VPort:

Example:
Command: **rename**
Block/Dimstyle/LAyer/LType/Style/Ucs/VIew/VPort: **B**
Old block name: **diode-20**
New block name: **diode-02**

COMMAND OPTIONS
Block   Change the name of a block.
Dimstyle   Change the name of a dimension style.
LAyer   Change the name of a layer.
LType   Change the name of a linetype.
Style   Change the name of a text style.
Ucs   Change the name of a UCS configuration.
VIew   Change the name of a view configuration.
VPort   Change the name of a viewport configuration.

RELATED AUTOCAD COMMANDS
- **DdLModes** Changes layer names via a dialogue box.
- **DdRename** Dialogue box for renaming.
- **DdUcs** Changes UCS configuration names via a dialogue box.

RELATED SYSTEM VARIABLES
- **CeLayer** Name of current layer.
- **CeLtype** Name of current linetype.
- **DimStyle** Name of current dimension style.
- **InsName** Name of current block.
- **TextStyle** Name of current text style.
- **UcsName** Name of current UCS view.

TIP
- When you have to rename the same objects (such as changing the names of layers to match a client's requirements) in many drawing files, use a script file to automate the procedure.

Regen

Rel. 1

Regenerates all viewports to update the drawing.

Command	Alt+	Menu bar	Alias	GC Alias
regen	V,G	[View] [Regen]	rg	...

Command: **regen**
Regenerating drawing.

COMMAND OPTION
[Ctrl]+C   Cancels the regeneration.

RELATED AUTOCAD COMMAND
- **Redraw**   Quickly cleans up all viewports.

RELATED SYSTEM VARIABLE
- **RegenMode** Current setting of **RegenAuto**:
  - 0 Off.
  - 1 On (default).

TIPS
- Some commands automatically force a regeneration of the screen; other commands queue the regen.

- To save on regeneration time:
  - Freeze layers you are not working with.
  - Use **QText** to turn text into rectangles.
  - Place hatching last on its own layer, and freeze that layer.

'Redraw   Rel. 1

Redraws all viewports to clean up the screen.

Command	Alt+	Menu bar	Alias	GC Alias
'redraw	V,R	[View] [Redraw]	r	rd

Command: `redraw`

Before redraw:

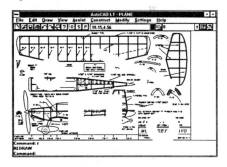

After redraw:

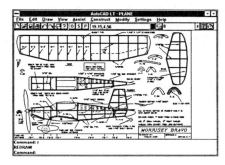

COMMAND OPTION
[Ctrl]+C   Cancel the redraw.

RELATED AUTOCAD COMMAND
- **Regen**   Regenerates all viewports.

RELATED SYSTEM VARIABLE
- **SortEnts** Controls the order of redrawing objects:
  - 0 Sorted by order in the drawing database.
  - 1 Sorted for object selection.
  - 2 Sorted for object snap.
  - 4 Sorted for redraw.
  - 8 Sorted for creating slides.
  - 16 Sorted for regenerations.
  - 32 Sorted for plotting.
  - 64 Sorted for PostScript plotting.

TIPS
- Use **Redraw** to clean up the screen after a lot of editing; some commands automatically redraw the screen when they are done.

- **Redraw** does not affect objects on frozen layers.

 # Redo

Rel. 1

Reverses the effect of the most recent **Undo** and **U** command.

Command	Alt+	Menu bar	Alias	GC Alias
redo	E,R	[Edit]	re	uu
		[Redo]		

Command: **redo**

COMMAND OPTIONS
None.

RELATED AUTOCAD COMMAND
- **Undo**   Undoes the most recent series of AutoCAD commands.

RELATED SYSTEM VARIABLE
- **UndoCtl**   Determines the state of the **Undo** command.

TIP
- The **Redo** command is limited to undoing a single undo, while the **Undo** and **U** commands undo operations all the way back to the beginning of the editing session.

 # Rectang

Rel. 1

Draws a rectangle out of a polyline.

Command	Alt+	Menu bar	Alias	GC Alias
rectang	D,G	[Draw] [Rectangle]	rc	re

Command: **rectangle**
First corner: **[pick]**
Other corner: **[pick]**

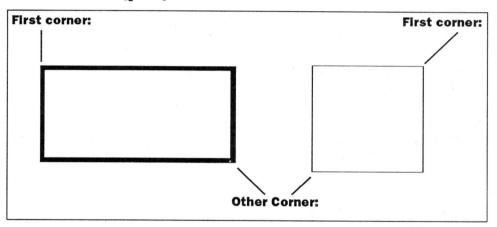

COMMAND OPTIONS
None.

RELATED AUTOCAD COMMANDS
- **Donut**   Draws solid-filled circles with a polyline.
- **Ellipse**   Draws ellipsis with a polyline.
- **PEdit**   Edit polylines, include rectangles.
- **PLine**   Draw polylines and polyline arcs.
- **Polygon**   Draw a polygon (3 to 1,024 sides) from a polyline.

TIPS

■ Rectangles are drawn from polylines; use the **PEdit** command to change the rectangle, such as the width of the polyline.

■ The pick point determines the location of rectangle's first vertex; rectangles are drawn counterclockwise.

■ Use the **Snap** command and object snap modes to precisely place the rectangle.

■ Use object snap mode **INTersection** to snap to the rectangle's vertices.

Quit

Rel. 1

Exits AutoCAD without saving changes to the drawing from the most recent **Save** or **End** command.

Command	Alt+	Menu bar	Aliases	GC Alias
quit	F,X	[File]	exit	...
		[Exit]	et	
			q	

Command: **quit**
Displays dialogue box:

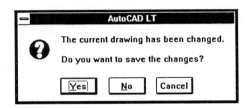

COMMAND OPTIONS
Yes — Save changes before leaving AutoCAD.
No — Don't save changes; exit AutoCAD.
Cancel — Don't quit AutoCAD; return to drawing.

RELATED AUTOCAD COMMANDS
- End — Saves the drawing and exits AutoCAD.
- SaveAs — Saves the drawing by another name or to another subdirectory or drive.

RELATED SYSTEM VARIABLES
None.

RELATED FILES
- *.Dwg — AutoCAD drawing files.
- *.Bak — Backup file.
- *.Bk1 — Additional backup files.

TIPS
- How to make changes to a drawing, yet preserve its original format:
    1. Use the **SaveAs** command to save the drawing by another name.
    2. Use the **Quit** command to preserve the drawing in its original state.

- How to use the DOS **Rename** command to recover the previous version when you accidently save over a drawing:
    1. Rename the most recently saved drawing filename.
    2. Then, rename the backup file (with BAK extension) to the DWG extension.

QText

Rel. 1

Displays all text as rectangular boxes (*short for Quick TEXT*).

Command	Alt+	Menu bar	Alias	GC Alias
qtext	S,D,A	[Settings] [Drawing Aids] [Quick Text]	qt	tv

Command: **qtext**
ON/OFF <Off>: **on**

COMMAND OPTIONS
ON   Turns quick text on, after the next **Regen** command.
OFF   Turns quick text off, after the next command.

RELATED AUTOCAD COMMANDS
- **DdRModes** Toggle quick text via dialogue box.
- **Regen** Regenerates the screen; makes quick text take effect.

RELATED SYSTEM VARIABLE
- **QTextMode** Holds the current state of quick text mode.

TIPS
- The length of a **QText** box does not necessarily match the actual length of text.

- To find invisible text (such as text made of spaces), turn on **QText**, thaw all layers, and **Zoom** to extents.

QSave

Rel. 1

Saves the current drawing without requesting a filename (*short for Quick SAVE*).

Command	Alt+	Menu bar	Alias	GC Alias
qsave	F,A	[File] [Save]	...	...

Command: **qsave**

COMMAND OPTIONS
None.

RELATED AUTOCAD COMMANDS
- End   Saves the drawing, without requesting a filename, and ends AutoCAD.
- Save   Saves drawing, after requesting the filename.
- SaveAs   Saves the drawing by a different filename.

RELATED SYSTEM VARIABLES
- DwgName   Current drawing filename (default: "UNNAMED").
- DwgTitled Status of drawing's filename:
  - 0 Name is "UNNAMED".
  - 1 Name is other than UNNAMED.
- DwgWrite   Drawing's read-write status:
  - 0 Read-only.
  - 1 Read-write.

TIPS
- When the drawing is not named, the **QSave** command requests a filename.

- When the drawing file, its subdirectory, or drive (such as a CD-ROM drive) are marked "read-only," use the **SaveAs** command to save the drawing to another filename, subdirectory, or drive.

Purge

Rel. 1

Removes unused named objects from the drawing: blocks, dimension styles, layers, linetypes, shapes, text styles, application id tables, and multiline styles.

Command	Alt+	Menu bar	Alias	GC Alias
purge	M,U	[Modify]	pr	cx
		[Purge]		du

```
Command: purge
Purge unused Blocks/Dimstyles/LAyers/LTypes/STyles/All: A
```

Sample response:
```
No unreferenced blocks found.
Purge layer DOORWINS? <N> y
Purge layer TEXT? <N> y
Purge linetype CENTER? <N> y
Purge linetype CENTER2? <N> y
No unreferenced text styles found.
```

COMMAND OPTIONS

Blocks	Named but unused blocks.
Dimstyles	Unused dimension styles.
LAyers	Unused layers.
LTypes	Unused linetypes.
STyles	Unused text styles.
All	Purge drawing of all five named objects, if necessary.

RELATED AUTOCAD COMMANDS

- **End** — Two **End** commands in a row can remove spurious information from a drawing.
- **WBlock** — Writes the current drawing to disk (with the * option) and removes spurious information from the drawing.

TIPS

- The **Purge** must be used as the first command used after a drawing is loaded. Otherwise, AutoCAD complains, "The PURGE command cannot be used now."

- It may be necessary to use the **Purge** command several times; follow each purge with the **End** command, then **Open** the drawing and **Purge** again. Repeat until **Purge** reports nothing to purge.

TIPS

- Use paper space to lay out mutiple views of a single drawing.

- Paper space is known as "drawing composition" in some other CAD packages.

- When a drawing is in paper space, AutoCAD displays 'P' on the status line and the paper space icon:

QUICK START: Enabing paper space.

Entering paper space for the first time can be a mystifying experience, since your drawing literally disappears. Here are the steps you take:

1. Turn Tilemode off:
   Command: **tilemode 0**

2. Enter paper space:
   Command: **pspace**

3. Although the drawing area goes blank, don't worry: your drawing has not been erased. To see your drawing, you need to create at least one viewport:
   Command **mview fit**
   Your drawing reappears!

4. Now switch back to model space:
   Command: **mspace**

5. Use the **Zoom** and **Pan** command to make the drawing smaller or larger within the paper space viewport.

6. Switch back to paper space with the **PS** command alias. Create a few more viewports by picking points with the **MView** command. Try overlapping a couple of viewports.

7. Switch back to model space with **MS** and set different zoom levels for each viewport.

8. Switch back to paper space with **PS**. Now use the **Move** and **Stretch** commands to change the position and size of the paper space viewports.

9. Draw a title border around all the viewports. Some other paper space-related command to experiment with are: **VpLayer, Zoom XP,** and **PsLtScale.**

P PSpace

Rel. 1

Switches from model space to paper space (*short for Paper SPACE*).

Command	Alt+	Toolbar	Alias	GC Alias
pspace	...	P	ps	...

Command: **pspace**

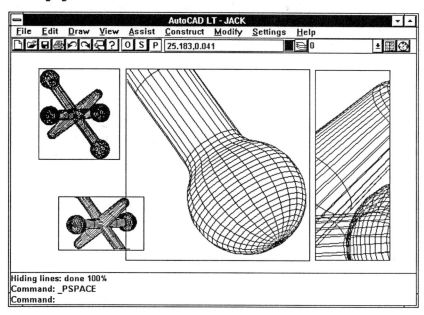

COMMAND OPTIONS
None.

RELATED AUTOCAD COMMANDS
- **MSpace**   Switches from paper space to model space.
- **MView**   Creates viewports in paper space.
- **UcsIcon** Toggles display of paper space icon.
- **Zoom**   The XP option scales paper space relative to model space.

RELATED SYSTEM VARIABLES
- **MaxActVp** Maximum number of viewports displaying an image.
- **TileMode** Must equal 0 for paper space to work.
- **PsLtScale** Linetype scale relative to paper space.

TIPS

■ The best size to select is as close to the final image size as possible; too small an image become coarse; too large an image creates a large file on disk.

■ The ? option does not work.

■ The 'screen preview image' is only used for screen display purposes, since graphics software generally cannot display PostScript graphic files.

■ Although Autodesk recommends using the smallest screen preview image size of 128x128, the largest preview image (512x512) has a minimal effect on file size and screen display time.

■ The screen preview image size has no effect on the quality of the PostScript output.

■ If you're not sure which screen preview format to use, select TIFF.

■ When PostScript fonts are incorrectly printed, comment out (using the ; semi-colon prefix) all entries for the fonts in the Aclt.Psf file. This causes **PsOut** to output the fonts as unfilled outlines. The drawback is a larger EPS file.

PsOut *cont'd*

```
A1          22.40       32.20
A0          32.20       45.90
USER         7.50       10.50
Enter the Size or Width,Height (in Inches) <USER>:
Effective plotting area: ww by hh high
```

COMMAND OPTIONS
What to export   Export all objects within:
   Display   The currently displayed view of the current viewport (default).
   Extents   The drawing extents.
   Limits   The drawing limits.
   View   A named view; AutoCAD prompts, 'View name: '.
   Window   A windowed selection; AutoCAD prompts, 'First corner: ' and 'Other corner:'.

Include a screen preview image in the file?
   None   No preview image (default).
   ESPI   Encapsulated PostScript raster format.
   TIFF   Tagged Image File Format.

Screen preview image size?
   128   128x128 pixels (default).
   256   256x256 pixels.
   512   512x512 pixels.

Size units   Select imperial or metric units:
   Inches   Imperial units (default).
   Millimeters Metric units.

RELATED AUTOCAD COMMANDS
- **SaveDIB**   Export drawing in Windows bitmap (raster) format.
- **Plot**   Exports dawing in HPGL vector and four raster formats.
- **WmfOut**   Export drawing in Windows metafile format.

RELATED SYSTEM VARIABLE
- **PsProlog**   Specify the PostScript prologue information.

RELATED FILES
- ***.Eps**   Extension of file produced by **PsOut**.
- ***.Pfm**   PostScript font definition files used by **PsOut**.
- **Aclt.Psf**   PostScript font substitution map, ISO font substitution procedures, and PostScript figure inclusion routines.

PsOut

Rel. 1

Exports the current drawing as an encapsulated PostScript file.

Command	Alt+	Menu bar	Alias	GC Alias
psout	F,I,T	[File] [Import/Export] [PostScript Out]	pu	...

Command: **psout**
Specify filename in dialogue box:

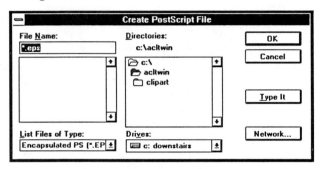

```
What to export -- Display, Extents, Limits, View or Windows
    <D>: [Enter]
Include a screen preview image in the file? (None/EPSI/TIFF)
    <None>: TIFF
Screen preview image size (128x128 is standard)? (128/256/
    512)<128>: 512
Size units (Inches or Millimeters) <Inches>: [Enter]
Specify scale by entering:
Output inches = Drawing Units or Fit or ? <Fit>: [Enter]
Standard values for output size
Size      Width          Height
A          8.00           10.50
B         10.00           16.00
C         16.00           21.00
D         21.00           33.00
E         33.00           43.00
F         28.00           40.00
G         11.00           90.00
H         28.00          143.00
J         34.00          176.00
K         40.00          143.00
A4         7.80           11.20
A3        10.70           15.60
A2        15.60           22.40
```

Preferences *cont'd*

Font   Select fonts for AutoCAD graphics and text screen; displays dialogue box:

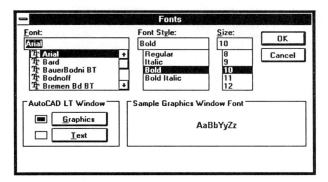

RELATED FILES
- *.Dwg   Prototype drawing files.
- *.Dwk   Locked drawing files.
- *.Mmu, *.Mnx Menu files (source and compiled).
- Aclt.Ini   Stores settings from **Preferences** command.
- Auto.Sv$ Automatic backup file.

Environment Names of subdirectories and menu files:
  Support Dirs Search path for subdirectories containing support files.
  Menu File Name of menu file (default = aclt.mnu).
  Browse Look for subdirectory names via dialogue box:

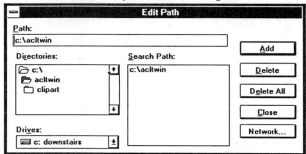

User Name The name used by the **About** and **RevDate** commands.
Temporary Files Location to store AutoCAD's temporary files; displays dialogue box:

Use Drawing Directory
    Store temporary files in current subdirectory.
Directory Name of subdirectory to store temporary files.
Browse Search for subdirectory name.

Color Specify colors for AutoCAD window; displays dialogue box:

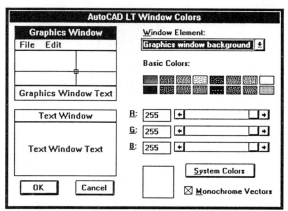

Preferences

Rel. 1

Lets you set user preferences.

Command	Alt+	Menu bar	Alias	GC Alias
preferences	F,F	[File] [Preferences]	pf	...

Command: **preferences**
Displays dialogue boxes.

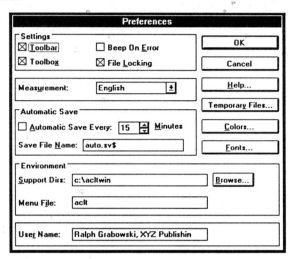

COMMAND OPTIONS

Settings   User interface options:
  Toolbar   Toggle display of the toolbar (on by default).
  Toolbox   Toggle display of the toolbox (on by default).
  Beep on Error Sound a beep when an error occurs (off by default).
  File Locking   Lock drawing file (*.Dwk) being used by AutoCAD (on by default).

Measurement Choose between imperial (default) or metric measurements.

Automatic Save Automatic file saving of current drawing options:
  Automatic Save Eevery
    Toggle automatic save (off by default).
  Minutes   Amount of time between automatic save:
- **Minimum** 1 minute.
- **Default**   15 minutes.
- **Maximum** 600 minutes (10 hours).

  Save File Name Name of file to automatically save to (default = auto.sv$).

218 ■ The Illustrated AutoCAD LT for Windows Quick Reference

RELATED SYSTEM VARIABLE

- **PolySides**   Most-recently specified number of sides; default is 4.

TIPS

- Polygons are drawn from polylines; use the **PEdit** command to change the polygon, such as the width of the polyline.

- The pick point determines the location of polygon's first vertex; polygons are drawn counterclockwise.

- Use the system variable **PolySides** to preset the default number of polygon sides.

- Use the **Snap** command to precisely place the polygon.

- Use object snap mode **INTersection** to snap to the polygon's vertices.

Polygon

Rel. 1

Draws a 2D polygon of between three to 1,024 sides.

Command	Alt+	Menu bar	Alias	GC Alias
polygon	D,Y	[Draw] [Polygon]	pg	rp

```
Command: polygon
Number of sides <4>:
Edge/<Center of polygon>: [pick]
Inscribed in circle/Circumscribed about circle (I/C): I
Radius of circle: [pick]
```

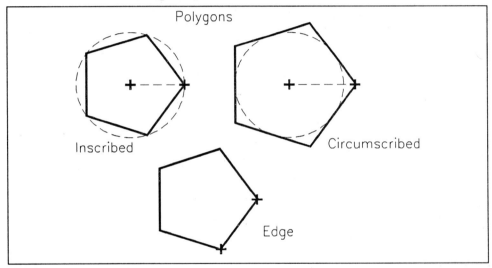

COMMAND OPTIONS

<Center of polygon>
    Indicate the center point of the polygon; then:
  C   Fit the polygon outside of a circle; or
  I   Fit the polygon inside a circle.

Edge   Draw the polygon based on the length of one edge.

RELATED AUTOCAD COMMANDS

- **Donut**   Draws solid-filled circles with a polyline.
- **Ellipse**   Draws ellipsis with a polyline.
- **PEdit**   Edit polylines, include polygons.
- **PLine**   Draw polylines and polyline arcs.
- **Rectang**   Draw a rectangle from a polyline.

 # Point

Rel. 1

Draws a 3D point.

Command	Alt+	Menu bar	Alias	GC Alias
point	D,O	[Draw] [Point]	pt	po

Command: **point**
Point: **[pick]**

COMMAND OPTIONS
None.

RELATED AUTOCAD COMMAND
- **DdPType**   Dialogue box for selecting **PsMode** and **PdSize**.

RELATED SYSTEM VARIABLES
- **PdMode**   Determines the look of a point:

```
   0    1    2    3    4
   .    |    +    ×    |

  32   33   34   35   36
  ⊙    ○    ⊕    ⊗    ⊙

  64   65   66   67   68
  ⊡    □    ⊞    ⊠    ⊡

  96   97   98   99   100
  ⊙    ◌    ⊕    ⊠    ⊙
```

- **PdSize**   Determines the size of a point:
  - 0 Point is 5% of height of **ScreenSize** system variable.
  - 1 No display.
  - -10 Ten percent of viewport size.
  - 10 Ten pixels in size.

TIPS
- The size and shape of the point is determined by **PdSize** and **PdMode**; changing these values alters the look and size of all points in the drawing with the next regeneration.

- Entering only x,y-coordinate places the point at a z-coordinate of the current elevation; setting **Thickness** to a value draws the point as a line in 3D space.

- Prefix the coordinate with * (asterisk) to place a point in the WCS, rather than the current UCS.

- Use the object snap mode **NODe** to snap to a point.

Plot *cont'd*

Full   Full plot preview; displays viewport:

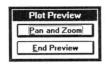

RELATED AUTOCAD COMMAND
- **PsOut**   Saves a drawing in EPS format.

RELATED SYSTEM VARIABLES
- **CmdDia**   Determines the **Plot** command's interface:
               0 Command-line interface (compatible with script files).
               1 Dialogue box interface.
- **PlotId**   Currently selected plotter number.
- **Plotter** Currently selected plotter name.

RELATED DOS VARIABLE
- **-p**   Starts AutoCAD LT with the **Plot** command:

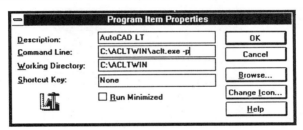

RELATED FILES
- ***.PCP**   Plotter configuration parameter files.
- ***.PLT**   Plot files created with the **Plot** command.

TIPS
- In LT Release 2, use the **Plot** command to export the drawing in the following raster formats in three resolutions (640x480, 800x600, 1024x768):
    - **BMP**   Microsoft's BitMaP.
    - **GIF**   CompuServe's Graphics Interchange Format.
    - **PCX**   Z-Soft's PC Paintbrush.
    - **TIFF**   Adobe's Tagged Image File Format.

- The only vector format supported is HPGL; use the **DxfOut**, **PsOut**, and **WmfOut** commands for exporting the drawing in other vector formats.

- Use the **SaveDIB** command as an alternative to use the **Plot** command for creating a BMP file of the drawing.

User Previous Printer
    Toggle whether to use Window system printer (default) or configured plotter (when turned on).
Plot to File Plot drawing to file.
Size Specify size of plot; displays dialogue box:

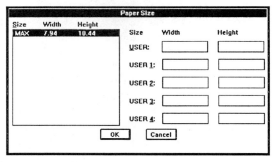

Rotation and origin
    Specify plot origin and rotation; displays dialogue box:

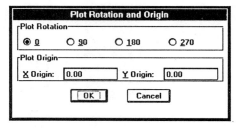

Scaled to fit Scale plot to fit paper size.

Preview Preview the plot.
  Partial   Quick plot preview; displays dialogue box:

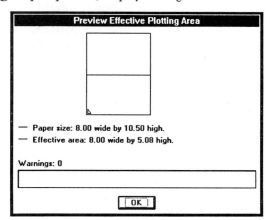

Plot *cont'd*

Pen assignments
    Assign pen numbers; displays dialogue box:

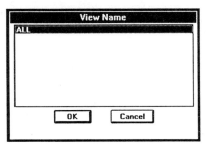

Display   Plot current display.
Extents   Plot drawing extents.
Limits   Plot drawing limits.
View   Plot named view; displays dialogue box if drawing contains saved views:

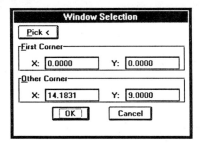

Window   Plot windowed area; displays dialogue box:

Hide lines   Remove hidden lines.
Adjust area fill   Adjust pen motion for filled areas.
Dithering   Convert colors to greypatterns for monochrome printer.

COMMAND OPTIONS

Device and default selection
Select and configure output devices; displays dialogue box:

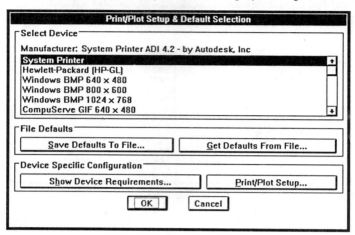

Show Device Requirements

```
           Show Device Requirements
Device Name:    HP LaserJet III
Output Port:    LPT1:
Driver Version: 3.10
Use the Control Panel to make permanent changes to
a printer's configuration.
                    [  OK  ]
```

Feature legend   Data varies, depending on device capabilities:

```
              Feature Legend
0. Solid Line
1. - - - - - - - - - - - - -
2. ........................
3. -.-.-.-.-.-.-.-.-.-.-.-.
4. -..-..-..-..-..-..-..-.
                    [  OK  ]
```

Plot

Rel. 1

Sends a copy of the drawing to the Windows system printer, or plotter via the serial or parallel ports; and converts drawing to raster format.

Command	Alt+	Menu bar	Alias	GC Alias
plot	F,P	[File] [Print/Plot]	pp	dp

Command: **plot**
Displays dialogue box:

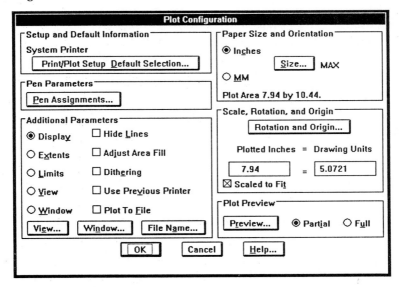

Halfwidth   Indicate the halfwidth of the polyline.
Length   Draw a polyline tangent the last segment.
Undo   Erase the last-drawn segment.
Width   Indicate the width of the polyline.
<Endpoint of line>:
                 Indicate the polyline's endpoint.

RELATED AUTOCAD COMMANDS
- **Boundary**   Draw a polyline boundary.
- **Donut**   Draws solid-filled circles as polyline arcs.
- **Ellipse**   Draws ellipses as polyline arcs.
- **Explode**   Reduces a polyline to lines and arcs with zero width.
- **Fillet**   Fillets polyline vertices with a radius.
- **PEdit**   Edits the polyline's vertices, widths, and smoothness.
- **Polygon**   Draws polygons as polylines of up to 1,024 sides.
- **Rectang**   Draws a rectangle out of a polyline.

RELATED SYSTEM VARIABLES
- **PlineGen**   Style of linetype generation:
                     0 Vertex to vertex (default).
                     1 End to end.
- **PlineWid**   Current width of polyline.

TIPS

■ Use the **Boundary** command to automatically outline a region; then use the **List** command to find its area.

■ If you cannot see a linetype on a polyline, change system variable **PlineGen** to 1; this regenerates the linetype from one end of the polyline to the other.

■ If the angle between a joined polyline and polyarc is less than 28 degrees, the transition is chamfered; at greater than 28 degrees, the transition is not chamfered.

■ Use the object snap mode **INTersection** to snap to the vertices of a polyline.

 # PLine

Rel. 1

Draws a complex 2D line made of straight and curved sections of constant and variable width; treated as a single object (*short for PolyLINE*).

Command	Alt+	Menu bar	Alias	GC Alias
pline	D,P	[Draw] [Polyline]	pl	...

```
Command: pline
From point: [pick]
Current line-width is 0.0
Arc/Close/Halfwidth/Length/Undo/Width/<Endpoint of line>:
```

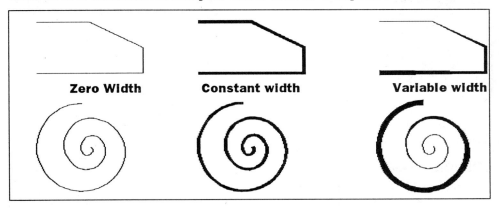

COMMAND OPTIONS

Arc	Displays the submenu for drawing arcs:
Angle	Indicate the included angle of the arc.
CEnter	Indicate the arc's center point.
CLose	Use an arc to close a polyline.
Direction	Indicate the arc's starting direction.
Halfwidth	Indicate the halfwidth of the arc.
Line	Switch back to the menu for drawing lines.
Radius	Indicate the arc's radius.
Second pt	For drawing a three-point arc.
Undo	Erase the last drawn arc segment.
Width	Indicate the width of the arc.
<Endpoint of arc>	
	Indicate the arc's endpoint.

 Close Close the polyline with a line segment.

Plan

Rel. 1

Displays the plan view of the WCS or the UCS.

Command	Alt+	Menu bar	Alias	GC Alias
plan	V,I,P	[View] [3D Plan View]	pv	...

Command: **plan**
<Current UCS>/Ucs/World: **W**
Regenerating drawing.

Example 3D view: *After using Plan World command:*

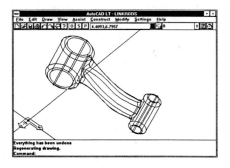 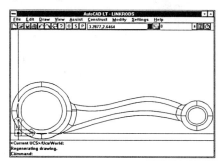

COMMAND OPTIONS
<Current UCS> Show the plan view of the current UCS.
Ucs   Show the plan view of a named UCS.
World   Show the plan view of the WCS.

RELATED AUTOCAD COMMANDS
- **UCS**   Creates new UCS views.
- **VPoint** Changes the viewpoint of 3D drawings.

RELATED SYSTEM VARIABLE
- **UcsFollow** Automatic plan view display for UCS or WCS.

TIPS
- Typing 'VPoint 0,0,0' is an alternative command to the **Plan World** command.

- The **Plan** command turns off perspective mode and clipping planes.

- **Plan** does not work in paper space.

- The **Plan** command is an excellent method for getting rid of the perspective mode created by the **DView** command.

PEdit *cont'd*

RELATED AUTOCAD COMMANDS
- **Break**   Breaks a 2D polyline at any position.
- **Chamfer**   Chamfers all vertices of a 2D polylines.
- **Fillet**   Fillets all vertices of a 2D polyline.
- **PLine**   Draws a 2D polyline.

RELATED SYSTEM VARIABLES
- **Splframe**   Determines visibility of a polyline spline frame.
- **SplineSegs** Number of lines used to draw a splined polyline.
- **SplineType** Determines B-spline smoothing for 2D and 3D polylines.
- **SurfType**   Determines the smoothing using the Smooth-surface option.

TIP
- During vertex editing, button #2 (left button on a two-button mouse) moves the X-marker to the next vertex.

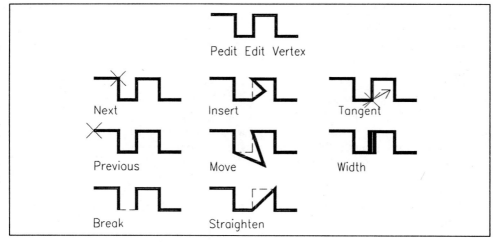

Go	Perform the break.
eXit	Exit the **Break** sub-submenu.
Insert	Insert another vertex.
Move	Relocate a vertex.
Next	Move the x-marker to the next vertex.
Previous	Move the x-marker to the previous vertex.
Regen	Regenerate the screen to show effect of **PEdit** commands.
Straighten	Draw a straight segment between two vertices:
\<Next\>	Move the x-marker to the next vertex.
Previous	Move the x-marker to the previous vertex.
Go	Perform the straightening.
eXit	Exit the **Straighten** sub-submenu.
Tangent	Show tangent to current vertex.
Width	Change the width of a segment.
\<eXit\>	Exit the **Edit-vertex** submenu.
Fit	Fit a curve to the tangent points of each vertex.
Ltype gen	Specify linetype generation style.
Join	Add other polylines to the current polyline.
Open	Open a closed polyline by removing the last segment.
Spline	Fit a splined curve along the polyline.
Undo	Undo the most-recent **PEdit** operation.
Width	Change the width of the entire polyline.
\<eXit\>	Exit the **PEdit** command.

 # PEdit   Rel. 1

Edits a 2D or 3D polyline — depending on which is picked (*short for Polyline EDIT*).

Command	Alt+	Menu bar	Alias	GC Alias
pedit	M,Y	[Modify] [Polyline Edit]	pe	cv

For 2D polylines:

```
Command: pedit
Select polyline: [pick a 2D polyline]
Close/Join/Width/Edit vertex/Fit/Spline/Decurve/Ltype gen
   /Undo/eXit <X>: e
Next/Previous/Break/Insert/Move/Regen/Straighten/Tangent
   /Width/eXit <N>: b
Next/Previous/Go/eXit <N>:
```

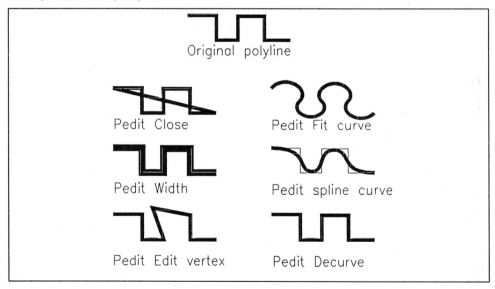

COMMAND OPTIONS

Close	Close an open polyline by joining the two endpoints with a single segment.
Decurve	Reverse the effects of a **Fit-curve** or **Spline-curve** operation.
Edit vertex	Edit individual vertices and segments (*see figure below*):
Break	Remove a segment or break the polyline at a vertex.
<Next>	Move the x-marker to the next vertex.
Previous	Move the x-marker to the previous vertex.

RELATED AUTOCAD COMMANDS
- **CopyClip**   Copies drawing to the Windows Clipboard.
- **CopyEmbed**   Copies to the Clipboard in OLE format.
- **CopyImage**   Copies windows area to the Clipboard in raster format.
- **CopyLink**   Copies current viewport to the Clipboard.
- **WmfIn**   Imports a WMF file in the drawing.
- **WmfOpts**   Specifies how an WMF file in placed in the drawing.
- **Paste**   Places Clipboard text in drawing via the **DText** command.

TIPS
- Contrary to the Autodesk documentation, the **PasteClip** does not place text, bitmaps, or multimedia objects in the drawing; there is no linking or embedding.

- The **PasteClip** command works only with WMF-format objects, also known as Picture format in the Windows Clipboard.

- AutoCAD LT Release 2 includes 39 WMF files in subdirectory \Acltwin\Clipart.

- The **PasteClip** command is not predictable in placing objects in the drawing.

PasteClip

Rel. 1

Place an WMF object in the drawing from Windows Clipboard (*short for PASTE from CLIPboard*).

Command	Alt+	Menu bar	Alias	GC Alias
pasteclip	E,A	[Edit] [Paste]	pc	...

Command: **pasteclip**
Command: _INSERT Block name (or ?): WMF0
 Insertion point: **[pick]**
X scale factor <1> / Corner / XYZ: **[Enter]**
 Y scale factor (default=X): **[Enter]**
 Rotation angle <0>: **[Enter]**

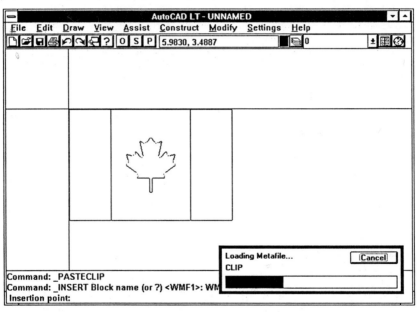

COMMAND OPTIONS

Insertion point   Indicate point to place picture.
X scale factor   Specify scale factor in x-direction.
Y scale factor   Specify scale factor in x-direction.
Rotation angle   Specify rotation angle in counterclockwise direction.

QUICK START: Pasting Text in the Drawing

How to use the **Paste Command** to paste text from the Windows Clipboard into the drawing:

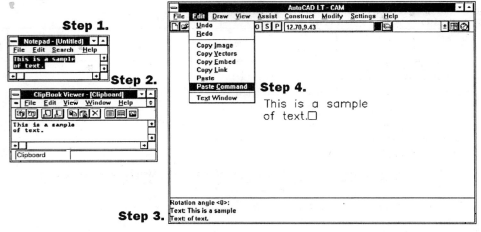

1. In a text editor, such as Windows Notepad, type the text or load a text file. Select the all the text with the **Edit | Select All** command from the menu bar.

2. With the **Edit | Copy** command, copy the selected text from Notepad to the Windows Clipboard.

3. Switch to AutoCAD and start the **DText** command, as follows:
   Command: **dtext**
   Justify/Style/<Start point>: **[pick]**
   Height <0.20>: **[Enter]**
   Rotation angle <0>: **[Enter]**
   Text:

4. Select the **Edit | Paste Command** command from the menu bar. The text stored in the Clipboard flows into the drawing, as follows:
   Text: **[Edit] [Paste Command]** This is a sample
   Text: of text. **[Enter]**
   Text: **[Enter]**

5. End the **DText** command by pressing the **[Enter]** key twice.

Paste Command

Rel. 1

Places text from Windows Clipboard in the drawing via the **Text** or **DText** commands.

Command	Alt+	Menu bar	Alias	GC Alias
...	E,C	[Edit] [Paste Command]	...	...

Example usage:
Command: **dtext**
Justify/Style/<Start point>: **[pick]**
Height <0.20>: **[Enter]**
Rotation angle <0>: **[Enter]**
Text: **[Edit] [Paste Command]** This is a sample
Text: of text.**[Enter]**
Text: **[Enter]**

COMMAND OPTIONS
None.

RELATED AUTOCAD COMMANDS
- **CopyClip** Copies drawing to the Windows Clipboard.
- **CopyEmbed** Copies to the Clipboard in OLE format.
- **CopyImage** Copies windows area to the Clipboard in raster format.
- **CopyLink** Copies current viewport to the Clipboard.
- **PasteClip** Places Clipboard WMF images in drawing.

TIPS
- Only text-format objects, also known as OEM Text, can be pasted with this command.

- The **Paste Command** command works only within the **DText** and **Text** commands:
  - **DText** Any length of text.
  - **Text** A single line of text.

- This command cannot be typed at the 'Command:' prompt, nor can it be used in toolbar macros; it can only be selected from the menu bar.

- The **Paste Command** is greyed out when the Clipboard contains non-text objects, such as a picture or bitmap.

- Click on the **Realtime** button (of the Aerial View window) to perform 'real-time' panning: the drawing pans as quickly as you move the mouse.

- You cannot use transparent pan during:
  - Paper space.
  - Perspective mode.
  - **VPoint** command.
  - **DView** command.
  - Another **Pan, View,** or **Zoom** command.

- The **DView** command has its own **Pan** option.

- When you press **[Enter]** to the 'Second point:' prompt, AutoCAD pans the drawing by the distance entered to the 'Displacement:' prompt.

 # 'Pan

Rel. 1

Moves the view in the current viewport to a different position.

Command	Alt+	Menu bar	Alias	GC Alias
'pan	V,P	[View] [Pan]	p	...

Command: **pan**
Displacement: [pick]
Second point: [pick]

Before panning: *After panning up:*

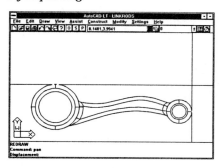 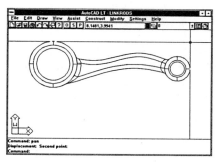

COMMAND OPTIONS
None.

RELATED AUTOCAD COMMANDS
- **DsViewer**   Aerial View pans in an independent window.
- **DView**   Pans during perspective mode.
- **View**   Saves and restores named views.
- **ViewRes**   Toggles whether pans are redrawn or regenerated.
- **Zoom**   The **Previous** option restores the view prior to panning.

RELATED SYSTEM VARIABLES
- **RegenAuto** Determines how regenerations are handled.
- **ViewCtr**   The x,y-coordinate of the view's center.
- **ViewDir**   View direction relative to UCS.
- **ViewSize**   Height of view in units.

TIPS
- You pan each viewport independently.

- You can use the **Pan** command transparently to start drawing an object in one area of the drawing, pan over, then continue drawing in another area of the drawing.

- **OsMode**   Current object snap mode settings:
  - 0   No object snap modes set.
  - 1   ENDpoint.
  - 2   MIDpoint.
  - 4   CENter.
  - 8   NODe.
  - 16   QUAdrant.
  - 32   INTersection.
  - 64   INSertion point.
  - 128   PERpendicular.
  - 256   TANgent.
  - 512   NEArest.
  - 1024   QUIck.

TIPS

- The **Aperture** command controls the snap area AutoCAD searches through.
- If AutoCAD finds no snap matching the current modes, the pick point is selected.

Ortho *cont'd*

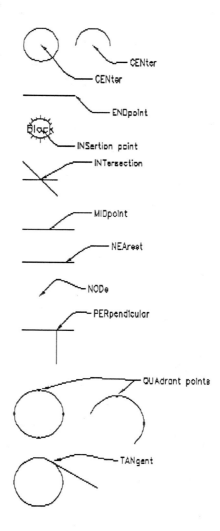

RELATED AUTOCAD COMMANDS
- **Aperture**   Controls the size of the object snap cursor up to 50 pixels.
- **DdOSnap**   Dialogue box for selecting object snap modes.

RELATED SYSTEM VARIABLES
- **Aperture**   Controls the size of the object snap cursor to any size.

'OSnap

Rel. 1

Sets and turns off object snap modes (*short for Object Snap*).

Command	Alt+	Mouse Button	Alias	GC Alias
'osnap	...	[Middle]	o	...

```
Command: osnap
Object snap modes:
```

COMMAND OPTIONS
As an abbreviation, enter only the first three letters:

- **ENDpoint:** Snap to end point of lines, polylines, traces, and arcs.
- **INTersection:** Snap to intersection of two entities.
- **MIDpoint:** Snap to middle point of lines and arcs.
- **PERpendicular:** Snap perpendicularly to entities.
- **CENter:** Snap to center point of arcs and circles.
- **TANgent:** Snap tangent to arcs and circles.
- **QUAdrant:** Snap the quadrant points of circles and arcs.
- **INSertion:** Snap to insertion point of blocks, shapes, and text.
- **NODe:** Snap to a point entity.
- **NEArest:** Snap to entity nearest to crosshairs.
- **NONe:** Temporarily turn off all object snap modes.
- **QUIck:** Snap to the first entity found in the database.
- **OFF** Turn off all object snap modes.

| O | 'Ortho   Rel. 1

Constrains drawing and editing commands to the vertical and horizontal directions only (*short for ORTHOgraphic*).

Command	Ctrl+	Function Key	Alias	GC Alias
'ortho	O	[F8]	or	or

Command: **ortho**
ON/OFF <Off>:

COMMAND OPTIONS
OFF   Turn ortho mode off.
ON   Turn ortho mode on.

RELATED AUTOCAD COMMAND
- **DdRModes** Toggles ortho mode via a dialogue box.

RELATED SYSTEM VARIABLE
- **OrthoMode** The current state of ortho mode.

TIPS
- Use ortho mode when you want to constrain your drawing and editing to right angles.

- Rotate the angle of ortho with the **Snap** command's **Rotate** option.

- In isoplane mode, ortho mode constrains the cursor to the current isoplane.

- Ortho mode is ignored when entering coordinates by keyboard and in perspective mode.

RELATED AUTOCAD COMMANDS
- **New**   Start a new drawing.
- **SaveAs**   Save drawing with a new name.

TIPS
- The **Open** command loads DWG drawing files for Release 12 and earlier.

- When a prior-to-Release 12 drawing is loaded, it is converted with the message, "Converting old drawing."

- Attempting to load a Release 13 drawing into AutoCAD LT results in the error message: "Was created by an incompatible version of AutoCAD."

- The **Open Drawing** dialogue box *appears* to allow AutoCAD LT to load the following filetypes: DWG, BMP, DCL, DXF, DXX, EPS, LIN, PAT, PCP, SHX, PFB, SCR, SLB, SLD, TXT, WMF, and XLG. However, it does not.

When you attempt to load any filetype other than DWG, and DXF — such as the WMF files supplied in subdirectory \Acltwin\Clipart — AutoCAD complains with the following dialogue box:

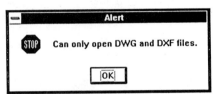

- The **Open** command does not load any file format into AutoCAD other than DWG and DXF. Instead, use these commands:
    - **Hatch**   To load PAT (hatch pattern) files.
    - **Linetype**   To load LIN (linetype) files.
    - **PasteClip**   To paste WMF (Windows metafile) files from Clipboard.
    - **Plot**   To load PCP (plot control parameters) files.
    - **Script**   To load SCR (script) files.
    - **Style**   To load PFB (PostScript font) files.
    - **VSlide**   To load SLD (slide) files.
    - **WmfIn**   To load WMF (Windows metafile) files.

 # Open

Rel. 1

Loads a drawing into AutoCAD.

Command	Alt+	Menu bar	Alias	GC Alias
open	F,O	[File] [Open]	op	ue

Command: **open**
Displays dialogue box:

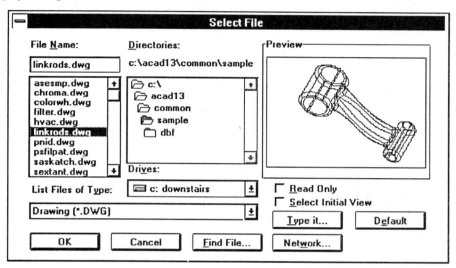

COMMAND OPTIONS

Default　　　Load current drawing again.
Pattern　　　Specify the filename pattern.
Read only mode　Displays drawing but you cannot edit it.
Select initial view
　　　　　　Select a named view from a dialogue box:

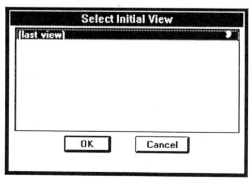

192 ■ The Illustrated AutoCAD LT for Windows Quick Reference

Oops   Rel. 1

Unerases the last-erased group of objects; returns the group of objects after the Block command.

Command	Alt+	Menu bar	Alias	GC Alias
oops	...	...	oo	ue

Command: **oops**

COMMAND OPTIONS
None.

RELATED AUTOCAD COMMAND
- **Undo**   Undoes the most-recent command.

RELATED SYSTEM VARIABLES
- *None.*

TIPS
- **Oops** only unerases the most-recently erased entity; use the **Undo** command to unerase earlier entities.

- Use **Oops** to bring back entities after turning them into a block with the **Block** and **WBlock** commands.

Offset

Rel. 1

Draws parallel lines, arcs, circles and polylines; repeats until cancelled.

Command	Alt+	Menu bar	Alias	GC Alias
offset	C,O	[Construct] [Offset]	of	sl

```
Command: offset
Offset distance or Through <Through>: t
Select object to offset: [pick]
Through point: [pick]
Select object to offset: [Esc]
```

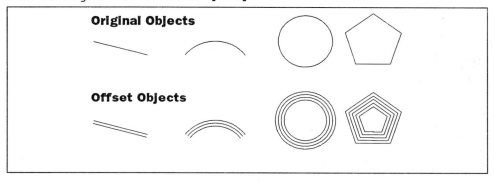

COMMAND OPTIONS
Through   Indicate the offset distance.
[Ctrl]+C   Exit the Offset command.

RELATED AUTOCAD COMMANDS
- Copy   Creates copies of a group of entities.
- Mirror Creates a mirror copy of a group of entities.

RELATED SYSTEM VARIABLE
- OffsetDist Current offset distance.

Archeng.Dwg, an example of a prototype drawing supplied with AutoCAD LT Release 2:

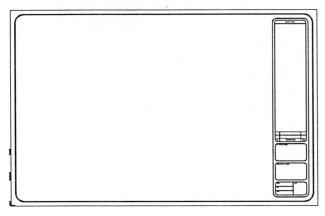

TIPS

- AutoCAD allows you to save your work before using the **New** command.

- Until you give the drawing a name, AutoCAD names the drawing Untitled.Dwg.

- The **Prototype** button lets you select a different prototype drawing from a dialogue box.

- The default prototype drawing is Acad.Dwg; edit and save Acad.Dwg to change the defaults in new drawings.

New *cont'd*

Connect Network Drive dialogue box:

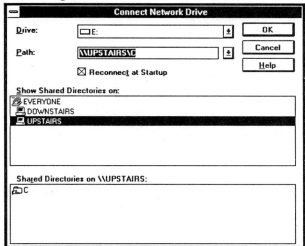

RELATED AUTOCAD COMMAND
- SaveAs   Saves the drawing by another name.

RELATED SYSTEM VARIABLES
- DbMod   Indicates whether drawing has changed since being loaded.
- DwgName   Name of current drawing.

RELATED FILES
North American prototype drawings:
- Acad.Dwg   The default prototype drawing.
- Ansi_a.Dwg   ANSI A-size (8.5" x 11").
- Ansi_b.Dwg   ANSI B-size (11"x17").
- Ansi_c.Dwg   ANSI C-size (17" x 22").
- Ansi_d.Dwg   ANSI D-size (22" x 34").
- Ansi_e.Dwg   ANSI E-size (34" x 44").
- Ansi_v.Dwg   ANSI vertical A-size (11" x 8.5").
- Archeng.Dwg   Architectural/engineering (see figure, next page).
- Gs24x36.Dwg   Generic D-size (24" x 36").

Metric prototype drawings:
- AcadIso.Dwg   The "generic" ISO prototype drawing.
- Iso_a0.Dwg   ISO A0-size.
- Iso_a1.Dwg   ISO A1-size.
- Iso_a2.Dwg   ISO A2-size.
- Iso_a3.Dwg   ISO A3-size.
- Iso_a4.Dwg   ISO A4-size (210mm x 297mm).

Open Drawing dialogue box:

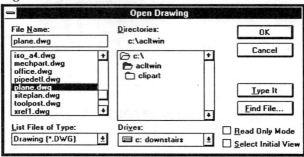

Find File dialogue box:

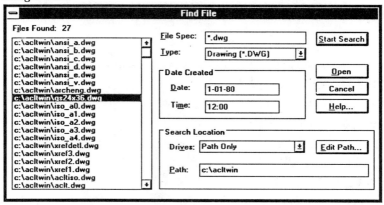

Edit Path dialogue box:

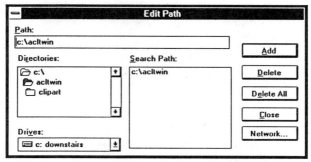

New *cont'd*

Custom Drawing Setup dialogue box:

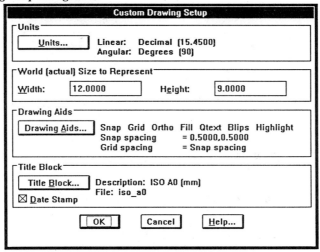

Title Block dialogue box:

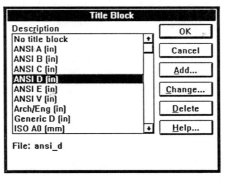

Add New Title Block dialogue box:

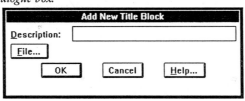

Change Title Block dialogue box:

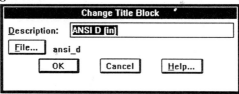

Prototype Drawing File dialogue box:

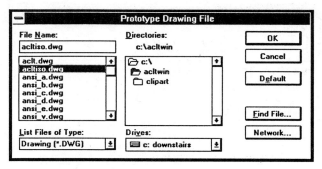

Quick Drawing Setup dialogue box:

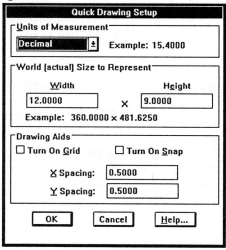

 # New Rel. 1

Name and start a new drawing.

Command	Alt+	Menu bar	Alias	GC Alias
new	F,N	[File] [New]	n	dx

Command: **new**
Displays dialogue box:

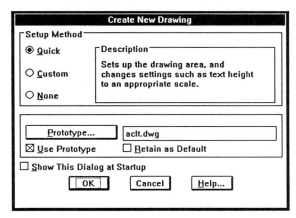

COMMAND OPTIONS

Setup Method Automatically sets up AutoCAD system variables for a new drawing:
    Quick   Sets some some parameters (see dialogue box, next page).
    Custom   Sets up the most parameters (see dialogue box, next page).
    None   Sets up no parameters.

Prototype   Specify the name of the prototype drawing (see dialogue box, next page).
Use Prototype Toggle whether to use a prototype drawing.
Retain as Default
                Specify whether to keep this prototype drawing for future new drawings.
Show This Dialog at Startup
                Toggle whether to be blessed by the **Create New Drawing** dialogue box in the future.

3   Displays the submenu of viewport orientations:
   Horizontal   Stack the three viewports.
   Vertical   Side-by-side viewports.
   Above   Two viewports above the third.
   Below   Two viewports below the third.
   Left   Two viewports to the left of the third.
   <Right>   Two viewports to the right of the third.

4   Displays the submenu of viewport orientations:
   Fit   Creates four same-size viewports to fit the screen.
   <First Point>   Indicate the area for the four viewports.

RELATED KEYSTROKE
- **[Ctrl]+V**   Switches to the next viewport.

RELATED AUTOCAD COMMANDS
- **MSpace**   Switches to model space.
- **PSpace**   Switches to paper space before creating viewports.
- **VpLayer**   Controls the visibility of layers in each viewport.
- **VPorts**   Creates tiled viewports in model space.
- **Zoom**   The **XP** option zooms a viewport relative to paper space.

RELATED SYSTEM VARIABLES
- **CvPort**   Current viewport.
- **MaxActVp**   Controls the maximum number of visible viewports:
                   1   Minimum.
                   16   Default.
                   32767   Maximum.
- **Tilemode** Controls the availability of overlapping viewports:
                   0   Off.
                   1   On.

TIPS
- Although system variable **MaxActVp** limits the number of simultaneously visible viewports, the **Plot** command plots all viewports.

- **Tilemode** must be set to zero to switch to paper space and use the **MSpace** command.

- **Snap, Grid, Hide, Shade,** etc., can be set separately in each viewport. The exception is the **Fill** command, which affects all viewports equally.

MView

Rel. 1

Creates and manipulates overlapping viewports (*short for Make VIEWports*).

Command	Alt+	Menu bar	Alias	GC Alias
mview	V,O	[View] [Viewports]	mv	vs

```
Command: mview
ON/OFF/Hideplot/Fit/2/3/4/Restore/<First Point>: [pick]
Other corner: [pick]
Regenerating drawing.
```

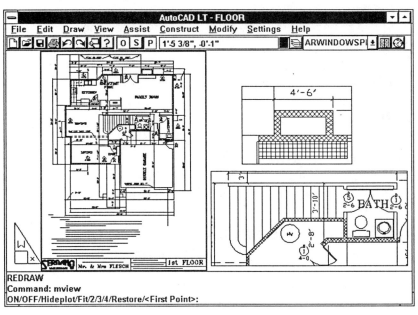

COMMAND OPTIONS

<First Point>	Indicate the first point of a single viewport.
Fit	Creates a single viewport that fits the screen.
Hideplot	Creates a hidden-line view during plotting and printing.
OFF	Turn a viewport off.
ON	Turn a viewport on.
Restore	Restore a saved viewport configuration.
2	Displays the submenu of viewport orientations:
Horizontal	Stack the two viewports.
<Vertical>	Side-by-side viewports.

Multiple

Rel. 1

A command modifier to automatically repeat commands.

Command	Alt+	Menu bar	Alias	GC Alias
multiple	...	...	...	...

Exampe usage:
Command: **multiple circle**
3P/2P/TTR/<Center point>: **[pick]**
Diameter/<Radius>: **[pick]**
circle 3P3P/2P/TTR/<Center point>: **[pick]**
Diameter/<Radius>: **[pick]**
circle 3P3P/2P/TTR/<Center point>: **[Ctrl]+C**

COMMAND OPTION

 [Ctrl]+C   Stop command from automatically repeating itself.

COMMAND INPUT OPTION

[Space]   Press the spacebar to repeat the previous command.

RELATED AUTOCAD COMMANDS

- Redo   Undoes an undo.
- U   Undoes the previous command; undoes a single multiple command at a time.

RELATED COMMAND MODIFIERS

- '   (*Apostrophe*) Allows use of certain transparent commands in another command.
- ~   (*Tilde*) Forces display of dialogue box.
- -   (*Dash*) Forces display of prompts on command line.
- tk   (*Tracking*) Visual entry of point filters.
- ↺   Direct distance entry.

TIPS

- **Multiple** is not a command but a command modifier; it does nothing on its own.
- **Multiple** only repeats the command name; it does not repeat command options.
- Some commands automatically repeat, including **Point** and **Donut**.

MSpace

Rel. 1

Switches the drawing from paper space to model space (*short for Model SPACE*).

Command	Alt+	Menu bar	Alias	GC Alias
mspace	...	...	ms	...

Command: **mspace**

COMMAND OPTIONS
None.

RELATED AUTOCAD COMMANDS
- **MView**   Creates viewports in paper space.
- **PSpace** Switches from model space to paper space.
- **VpLayer** Sets independent visibility of layers.

RELATED SYSTEM VARIABLES
- **MaxActVp** Maximum number of viewports with visible entities (default = 16).
- **PsLtScale** Linetype scale relative to paper space.
- **Tilemode** The current setting of tilemode:
  - 0 Off.
  - 1 On (the default).

TIPS
- **Tilemode** must be set to zero to switch to paper space and use the **MSpace** command.

- To switch from paper space back to model space, at least one viewport must be active; create the viewport with the **MView** command.

- Objects in the current selection set are ignored if they were not collected in the current space.

- AutoCAD clears the selection set when moving between paper and model space.

MSlide

Rel. 1

Save the current view as an SLD-format slide file on disk (*short for Make SLIDE*).

Command	Alt+	Menu bar	Alias	GC Alias
mslide	F,I,S	[File] [Import/Export] [Make Slide]	ml	...

```
Command: mslide
Slide file:
```

COMMAND OPTIONS
None.

RELATED AUTOCAD COMMANDS
- **Save** Saves the current drawing as a DWG-format drawing file.
- **VSlide** Displays an SLD-format slide file in AutoCAD.

RELATED AUTODESK PROGRAM
- **Slidelib.Exe** Compiles a group of slides into an SLB-format slide library file.

Move

Rel. 1

Moves a group of entities to a new location.

Command	Alt+	Menu bar	Alias	GC Aliases
move	M,M	[Modify]	m	mv
		[Move]		wm

```
Command: move
Select objects: [pick]
Select objects: [Enter]
Base point or displacement: [pick]
Second point of displacement: [pick]
```

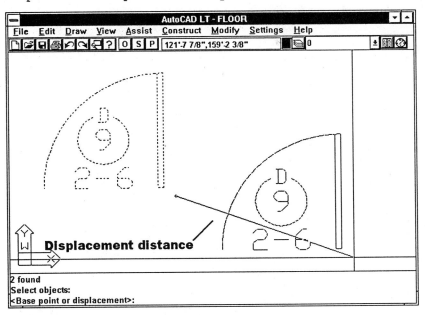

COMMAND OPTIONS
Base point Indicate the starting point for the move.
Displacement Indicate the distance to move.

RELATED AUTOCAD COMMANDS
- Copy   Makes a copy of selected objects.
- PEdit Moves the vertices of a polyline.

 # Mirror

Rel. 1

Creates a mirror copy of a group of entities in 2D space.

Command	Alt+	Menu bar	Alias	GC Alias
mirror	C,M	[Construct] [Mirror]	mi	mi

```
Command: mirror
Select objects: [pick]
Select objects: [Enter]
First point of mirror line: [pick]
Second point: [pick]
Delete old objects? <N>
```

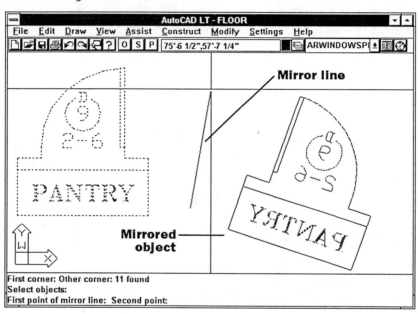

COMMAND OPTIONS
N   Do no delete selected objects.
Y   Delete selected objects.

RELATED AUTOCAD COMMAND
- Copy   Creates a non-mirrored copy of a group of entities.

RELATED SYSTEM VARIABLE
- MirrText   Determines whether text is mirrored by the **Mirror** command:
  0 Text is not mirrored about horizontal axis.
  1 Text is mirrored.

 # Measure

Rel. 2

Divides lines, arcs, circles, and polylines into equi-distant segments, placing a point or a block at each segment.

Command	Alt+	Menu bar	Alias	GC Alias
measure	C,S	[Construct] [Measure]	...	...

```
Command: measure
Select object to measure: [pick]
<Segment length>/Block: B
Block name to insert:
Align block with object? <Y>
Segment length:
```

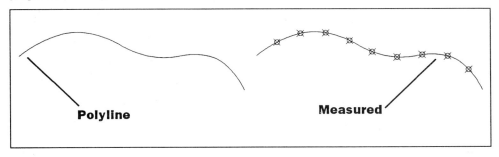

COMMAND OPTIONS
Block   Indicate the name of the block to use as a marker.
<Segment length>
     Indicate the distance between markers.

RELATED AUTOCAD COMMANDS
- Block   Creates blocks that can be used with **Measure**.
- Divide   Divides an entity into a number of segments.

RELATED SYSTEM VARIABLES
- PdMode   Controls the shape of a point.
- PdSize   Controls the size of a point.

TIPS
- You must define the block before it can be used with the **Measure** command.

- The **Measure** distance does not place a point or block at the beginning of the measured entity.

'LtScale

Rel. 1

Sets the scale factor of linetypes (*short for Line Type SCALE*).

Command	Alt+	Menu bar	Alias	GC Alias
'ltscale	S,N,N	[Settings]	lc	lz
		[Linetype Style]		
		[Linetype Scale]		
	S,N,P	[Settings]		
		[Linetype Scale]		
		[Pspace LT Scale]		

```
Command: ltscale
New scale factor <1.0000>:
Regenerating drawing.
```

Dashed **Dashed2** **DashedX2**

COMMAND OPTIONS
None.

RELATED AUTOCAD COMMAND
- **Linetype** Loads, creates, and sets the working linetype.

RELATED SYSTEM VARIABLES
- **LtScale** Contains the current linetype scale factor.
- **PlineGen** Controls how linetypes are generated for polylines.
- **PsLtScale** Linetype scale relative to paper space.

TIPS
- If the linetype scale is too large, the linetype appears solid.

- If the linetype scale is too small, the linetype appears as a solid line that redraws very slowly.

- In addition to setting the scale with the **LtScale** command, the Aclt.Lin and Acltiso.Lin files contain each linetype in three scales: normal, half-size, and double-size.

LogFileOn

Rel. 13

Opens Aclt.Log file and records 'Command:' prompt text to the file.

Command	Alt+	Menu bar	Alias	GC Alias
logfileon ...	...	...	...	...

Command: `logfileon`

COMMAND OPTIONS
None.

RELATED AUTOCAD COMMAND
- **LogFileOff** Turns off recording 'Command:' prompt text to file Aclt.Log.

RELATED FILE
- **Aclt.Log** Log file.

TIPS
- If log file recording is left on, it resumes when AutoCAD is next loaded.
- AutoCAD places a dashed line at the end of each log file session.

LogFileOff

Rel. 1

Closes the Aclt.Log file.

Command	Alt+	Menu bar	Alias	GC Alias
logfileoff ...	...	...	...	...

Command: `logfileoff`

COMMAND OPTIONS
None.

RELATED AUTOCAD COMMAND
- LogFileOn   Turns on recording 'Command:' prompt text to file Aclt.Log.

RELATED FILE
- Aclt.Log   Log file.

TIP
- AutoCAD places a dashed line at the end of each log file session.

List

Rel. 1

List information about selected entities in the drawing.

Command	Alt+	Menu bar	Alias	GC Alias
list	A,L	[Assist] [List]	ls	...

Command: **list**
Select objects: **[pick]**
Select objects: **[Enter]**

Example output:

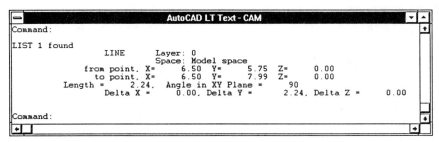

COMMAND OPTIONS
[Ctrl]+C   Cancel the list display.
[Ctrl]+S   Pause the display.
[Enter]   Continue the display.
[F2]   Return to graphics screen.

RELATED AUTOCAD COMMANDS
- **Area**   Calculates the area and perimeter of some entities.
- **Dist**   Calculates the 3D distance and angle between two points.
- **DdModify** Displays all known information about an entity and allows changes.

TIPS
- Use the **List** command as a faster alternative to using the **Dist** and **Area** commands for finding lengths and areas of objects.

- Data listed *conditionally* by the **List** command:
    - Object's color and linetype, if not set BYLAYER.
    - Thickness, if not 0.
    - The elevation is not listed and must be interpolated from the z-coordinate.
    - Extrusion direction, if different from z-axis of current UCS.

- Entity handles are described by hexadecimal numbers.

QUICK START: Create a Custom Linetype

To create a custom linetype on-the-fly:

1. Type the **Linetype Create** command:
   ```
   Command: linetype
   ?/Create/Load/Set: C
   ```

2. Name the linetype in three steps:
   - First, the linetype name:
     ```
     Name of linetype to create: [enter up to 31 characters]
     ```
   - Second, the LIN filename.
     Append linetype description to Acad.Lin or create new LIN file.
   - Third, describe the linetype:
     ```
     Descriptive text: [enter up to 47 characters]
     ```

3. Define the linetype pattern by using five codes:
   - Positive number for dashes; 0.5 is a dash 0.5 units long.
   - Negative number for gaps; -0.25 is a gap 0.25 units long.
   - Zero is for dots; 0 is a single dot.
   - "A" forces the linetype to align between two endpoints (linetypes start and stop with a dash).
   - Commas (,) separate values.

Example:
```
*DASHDOT,__ . __ . __ . __ . __ . __ . __ . __ . __ .
A,.5,-.25,0,-.25 [Enter]
```

4. Press **[Enter]** to end linetype definition.

5. Use the **Linetype Load** command to load pattern into drawing.
   ```
   Linetype to load: [type name]
   ```

6. Use the **Linetype Set** command to set the linetype.
   ```
   New object linetype (or ?) <>: [type name]
   ```
   Alternatively, use the **Change** command to change objects to the linetype.

Linetype *cont'd*

RELATED FILES

- The following standard linetypes are supplied with AutoCAD LT; their definitions are stored in files Aclt.Lin:

- A second linetype definition file, Acltiso.Lin, contains the same patterns but adjusted for ISO (metric) drawings.

'Linetype

Rel. 1

Loads linetype definitions into the drawing, creates new linetypes, and sets the working linetype.

Command	Alt+	Menu bar	Alias	GC Alias
'linetype	...	...	lt	lt

```
Command: linetype
?/Create/Load/Set:
```

COMMAND OPTIONS
Create   Create a new user-defined linetype.
Load   Load a linetype from an LIN linetype definition file.
Set   Set the working linetype.
?   List the linetypes loaded into the drawing.

RELATED AUTOCAD COMMANDS
- **Change**   Changes objects to a new linetype; changes linetype scale.
- **ChProp**   Changes entities to a new linetype.
- **DdEModes** Sets the working linetype via a dialogue box.
- **DdLModes** Sets the linetype for all entities on a layer.
- **LtScale**   Sets the scale of the linetype.
- **Rename**   Changes the name of the linetype.

RELATED SYSTEM VARIABLES
- **CeLtype**   The current linetype setting.
- **PsLtScale** Linetype scale relative to paper scale.
- **PlineGen** Controls how linetypes are generated for polylines.

TIPS
- The only linetype initially defined in an AutoCAD drawing is the CONTINUOUS linetype.

- Linetypes must be loaded from LIN definition files before being used in a drawing.

- When loading one or more linetypes, it is faster to load all linetypes. Then use the **Purge** command to remove linetype definitions not used by the drawing.

 # Line

Rel. 1

Draws straight 2D and 3D lines.

Command	Alt+	Menu bar	Alias	GC Alias
line	D,L	[Draw] [Line]	l	li

```
Command: line
From point: [pick]
To point: [pick]
To point: [Enter]
```

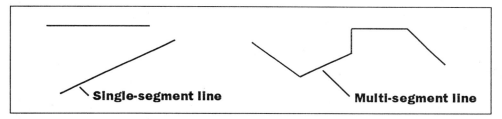

Single-segment line   Multi-segment line

COMMAND OPTIONS

 C   Close the line from the current point to the starting point.

U   Undo the last line drawn.

[Enter]   At the 'From point:' prompt, continues the line from the last endpoint; at the 'To point:' prompt, terminates the **Line** command.

RELATED AUTOCAD COMMAND
- **PLine**   Draws polylines and polyline arcs.

RELATED SYSTEM VARIABLES
- **Elevation**   Distance above (or below) the x,y-plane a line is drawn.
- **Lastpoint**   Last-entered coordinate triple.
- **Thickness**   Determines thickness of the line.

TIPS

- To draw a 2D line, enter x,y-coordinate pairs; the z-coordinate takes on the value of the Elevation system variable.

- To draw a 3D line, enter x,y,z-coordinate triples.

- When system variable **Thickness** is not zero, the line has thickness, which makes it a plane.

RELATED SYSTEM VARIABLES
- **LimCheck** Toggle for limit's drawing check:
    0 Off (default).
    1 On.
- **LimMin** Lower-right 2D coordinates of current limits.
- **LimMax** Upper-left 2D coordinates of current limits.

TIPS
- Use the **Limits** command to define the extents of grid markings.

- The limits determine the extents displayed by **Zoom All** command.

- If limits checking is turned on, AutoCAD will complain with an outside-limits error. Use this feature to prevent drawing outside of the drawing extents.

- There are no limits for the z-direction.

- Model space and paper space have separate limits.

'Limits

Rel. 1

Defines the 2D limits in the WCS for the grid markings and the Zoom All command.

Command	Alt+	Menu bar	Alias	GC Alias
'limits	S,W,L	[Settings] [Drawing] [Limits]	lm	ls

```
Command: limits
Reset Model space limits:
ON/OFF/<Lower left corner> <0.0000,0.0000>: [pick]
Upper right corner <12.0000,9.0000>: [pick]
```

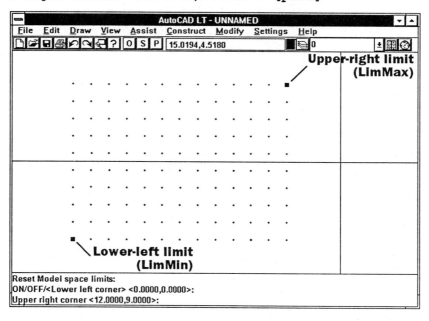

COMMAND OPTIONS
OFF   Turn limits checking off.
ON   Turn limits checking on.
[Enter]   Retain limits values.

RELATED AUTOCAD COMMANDS
- Grid   Grid dots are bounded by limits.
- Status   Lists the current drawing limits.
- Zoom   Zoom All displays the drawing's extents or limits.

COMMAND INPUT OPTIONS

*	Specify all layer names, such as:
	`Layer name(s) for color 3 (green): *`
* *and* ?	Specify several layer names, such as:
	`Layer name(s) for color 3 (green): ar*`
,	(Comma) Separates several layer names, such as:
	`Layer name(s) for color 3 (green): art,border`

RELATED AUTOCAD COMMANDS

- **Change**   Move objects to a different layer.
- **ChProp**   Move objects to a different layer.
- **DdLModes**   Use a dialogue box to control layers.
- **Purge**   Removed unused layers from the drawing.
- **Rename**   Rename layer names.
- **VpLayer**   Control the visibility of layers in paper space viewports.
- **XBind**   Bind in layers from other drawings.

RELATED SYSTEM VARIABLE

- **CeLayer**   Contains the name of the current layer.

TIPS

- Use layers to separate elements of a drawing.

- Every new drawing has just one layer, named '0'.

- Layer 0 cannot be purged.

- The maximum number of layers in a drawing is 32,000.

- A layer is created by simply giving it a name; the longest name is 31 characters long.

- Layer DimPts is a non-plotting layer.

- Use the XRef command to view layers of other drawing.

- Layer names that are xref'ed use the vertical bar (|) to separate drawing and layer names, such as:
    `dwgname|layername`

- Use the XBind command to bring in layers from other drawings.

- Layer names that have been xbind'ed use '0' to separate drawing and layer names, such as:
    `dwgname$0$layername`

- Place redraw-intensive objects, such as hatch and text, on a separate, frozen layer.

 # 'Layer

Rel. 1

Controls the creation and visibility of layers.

Command	Alt+	Menu bar	Alias	GC Aliases
'layer	...	...	la	yc
				yd
				yh
				yn

Command: `layer`
?/Make/Set/New/ON/OFF/Color/Ltype/Freeze/Thaw/LOck/Unlock:

All layers turned on: *Layer 'Dim' frozen:*

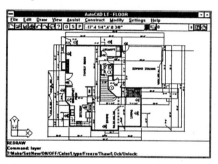 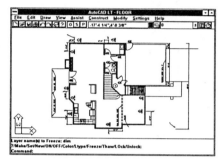

COMMAND OPTIONS

Color	Indicate the color for all entities drawn on the layer.
Freeze	Disable the display of the layer.
LOck	Lock the layer.
Ltype	Indicate the linetype for all entities drawn on the layer.
Make	Create a new layer and make it the working layer.
New	Create a new layer.
OFF	Turn the layer off.
ON	Turn the layer on.
Set	Make the layer the working layer.
Thaw	Un-freeze the layer.
Unlock	Unlock the layer.
?	List the names of layers created in the drawing.

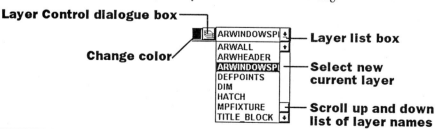

164 ■ The Illustrated AutoCAD LT for Windows Quick Reference

4. To create an isometric arc, place the isometric circle, then use the **Break** or **Trim** commands to erase the unwanted portion.

5. AutoCAD provides no explicit support for isometric text. To make text look correct on an isometric drawing, set up three styles with the following parameters:

- For text placed on the top isoplane:
    - **Style name:** "IsoTop"
    - **Oblique angle:** 0 degrees
    - **Text angle:** 0 degrees
- Text placed on the left isoplane:
    - **Style name:** "IsoLeft"
    - **Oblique angle:** -30 degrees
    - **Text angle:** -30 degrees
- Text for the right isoplane:
    - **Style name:** "IsoRight"
    - **Oblique angle:** 30 degrees
    - **Text angle:** 30 degrees

6. Similarly, AutoCAD does not support isometric dimensions. Use the dimension style feature to create three dimstyles that make dimensions look correct in an isometric drawing, as follows:

- Dimension style for the top isoplane:
    - **Text style:** IsoTop
    - **Dim Oblique:** -30 degrees
    - **Dim TRotate:** 0 degrees
- Dimstyle for the left isoplane:
    - **Text style:** IsoLeft
    - **Dim Oblique:** 30 degrees
    - **Dim TRotate:** -30 degrees
- Dimstyle for the right isoplane:
    - **Text style:** IsoRight
    - **Dim Oblique:** -30 degrees
    - **Dim TRotate:** 30 degrees

Isoplane *cont'd*

QUICK START: Using Isometric Mode

1. Switching to isometric mode is hidden away in the **Snap** command, as follows:
   Command: **snap**
   Snap spacing or ON/.../Style: **s**
   Standard/Isometric <S>: **i**

2. The grid and cursor are now in isometric mode. To switch between the three isometric drawing planes — left, top, right — you have three options, as follows:
   - The **Isometric** command:
       Command: **isoplane**
       Left/Top/Right/<Toggle>: **[Enter]**
       Current Isometric plane is: Left
   - Function key F5:
       Command: **[F5]** <Isoplane top>
   - Control key E:
       Command: **[Ctrl]+E** <Isoplane right>

3. The **Ellipse** command now includes the **Isocircle** option to help you draw isometric circles, as follows:
   Command: **ellipse**
   <Axis endpoint 1>/Center/Isocircle: **i**
   Center of circle: **[pick]**
   <Circle radius>/Diameter: **[pick]**

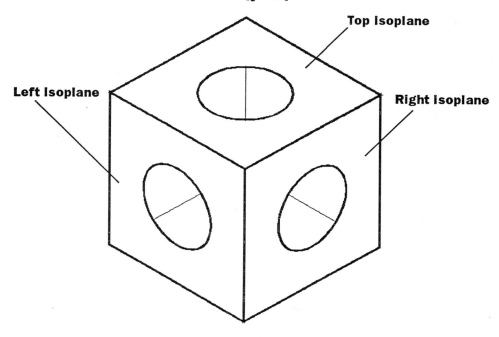

'Isoplane

Rel. 1

Switches the crosshairs between the three isometric drawing planes.

Command	Ctrl+	Menu bar	Function Key	Alias
'isoplane	E	[Settings]	[F5]	is
		[Drawing Aids]		
		[Isometric Snap/Grid]		

Command: **isoplane**
Left/Top/Right/<Toggle>: **[Enter]**
Current Isometric plane is: Left

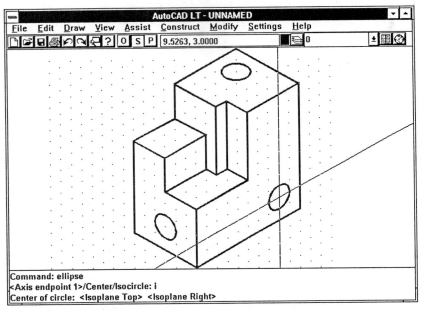

COMMAND OPTIONS
Left   Switch to the left isometric plane.
Top   Switch to the top isometric plane.
Right   Switch to the right isometric plane.
<Toggle>   Press **[Enter]** to switch to the next isometric plane in the order of: left, top, right.

RELATED AUTOCAD COMMANDS
- **DdRModes** Dialogue box for setting isometric mode and planes.
- **Ellipse** Draws isometric circles.
- **Snap** Turns on isometric drawing mode.

RELATED SYSTEM VARIABLE
- **SnapIsoPair** Contains the current isometric plane.

Insert *cont'd*

TIPS

- You can insert any other AutoCAD drawing into the current drawing.

- A "preset" scale factor or rotation means the dragged image is shown at that factor.

- Drawings are normally inserted as a block; prefix the filename with an * (asterisk) to insert the drawing as separate entities.

- Redefine an existing block by adding the suffix = (equal) after its name at the 'Block name:' prompt.

- Insert a mirrored block by supplying a negative x- or y-scale factor (such as 'X scale factor: -1'); AutoCAD converts negative z-scale factors into their absolute value (makes them always positive).

X scale factor Indicate the x-scale factor (default = 1); or:
  Corner Indicate the x- and y-scale factors by pointing on the screen:
  Other corner Pick other corner.
  XYZ Displays the x-, y- and z-scale submenu:
    X scale factor/Corner Specify the value of the y-scale (default = 1).
    Y scale factor Specify the value of the y-scale (default = x scale).
    Z scale factor Specify the value of the z-scale (default = x scale).

Rotation angle Specify the angle of rotation.

INPUT OPTIONS

- In response to the 'Block Name:' prompt, you can enter:
  - **?** Display a list of block names currently in the drawing.
  - **~** Display a dialogue box of blocks stored on disk.
  - ***** Prefix block's name with * to insert the block exploded, as in:
    Block name: ***blockname**
  - **=** Redefine existing block with a new block, as in:
    Block name: **oldname=newname**

- In response to the 'Insertion point:' prompt, you may enter:
  - **Scale** Specify x-, y-, z-scale factors.
  - **PScale** Preset the x-, y-, and z-scale factors.
  - **Xscale** Specify x-scale factor.
  - **PsScale** Preset x-scale factor.
  - **Yscale** Specify y-scale factor.
  - **PyScale** Preset y-scale factor.
  - **Zscale** Specify z-scale factor.
  - **PzScale** Preset the z-scale factor.
  - **Rotate** Specify the rotation angle.
  - **PRotate** Preset the rotation angle.

RELATED AUTOCAD COMMANDS

- **Block** Creates a block of a group of entities.
- **BMake** Creates a block using a dialogue box.
- **DdInsert** Dialogue box for inserting blocks.
- **Explode** Reduces an inserted block to its constituent entities.
- **Rename** Renames blocks.
- **WBlock** Writes blocks to disk.
- **XRef** Display drawings stored on disk in the drawing.

RELATED SYSTEM VARIABLE

- **InsBase** Name of most-recently inserted block.

 # Insert

Rel. 1

Inserts a previously-defined block (symbol) into the drawing.

Command	Alt+	Menu bar	Alias	GC Aliases
insert	...	...	in	ci
				cp

```
Command: insert
Block name (or ?):
Insertion point: [pick]
X scale factor <1> / Corner / XYZ: [Enter]
Y scale factor (default=X): [Enter]
Rotation angle <0>: [Enter]
```

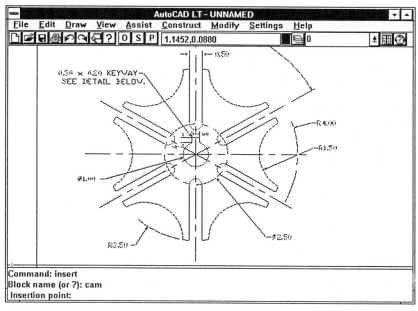

COMMAND OPTIONS

Block name   Indicate the name of the block to be inserted; o.:
   ?   List the names of blocks stored in the drawing.

Insertion point: Specify insertion point as x,y- or x,y,z-coordinate; or:
                     Supply a predefined scale and rotation angle (see input options).

'Id

Rel. 1

Identifies the 3D coordinates of a point (*short for IDentify*).

Command	Alt+	Menu bar	Alias	GC Alias
'id	A,I	[Assist] [ID Point]	...	...

```
Command: id
Point: [pick]
```

Example output:
```
X = 1278.0018    Y = 1541.5993    Z = 0.0000
```

INPUT OPTION
@   Using @ at the next prompt for coordinate input uses the value stored in system variable **LastPoint**.

RELATED AUTOCAD COMMANDS
- **List**   Lists information about a picked entity.
- **Point**   Draws a point.

RELATED SYSTEM VARIABLE
- **LastPoint**   The 3D coordinates of the last picked point.

TIPS
- The **Id** command stores the picked point in the **LastPoint** system variable. Access that value by entering '@' at the next prompt for a point value.

- Use the **Id** command to set the value of the **LastPoint** system variable, which can be used as relative coordinates in another command.

- The z-coordinate displayed by the **Id** command is the current elevation setting; if you use the **Id** command with an object snap, then the z-coordinate is the osnapped value.

- In a toolbar or toolbox macro, the **Id** command quickly label points on the screen.

Hide *cont'd*

RELATED SYSTEM VARIABLES
- *None.*

TIPS
- The **Hide** command considers the following entities as opaque:
    - Circle.
    - Solid and wide polyline.
    - The extrusion of any entity with thickness.

- Use the **MSlide** (or **SaveImg**) to save the hidden-line view as an SLD (or TIFF, TGA, or GIF) file. View the saved image with the **VSlide** (or **Replay**) command.

- The **Shade** command simulates hidden-line removal when **ShadEdge** system variable is set to 2.

- Freezing layers speeds up the hide process since **Hide** ignores those objects.

- **Hide** does not consider the visibility of text and attributes, unless the text is given thickness.

- To create a hidden-line view when plotting in paper space, select the **HidePlot** option of the **MView** command.

- Use the **Regen** command to return to the wireframe view.

Hide

Rel. 1

Removes hidden lines from 3D drawings.

Command	Alt+	Menu bar	Alias	GC Alias
hide	V,H	[View] [Hide]	hi	...
	V,S,H	[View] [Shade] [16 Color Hidden Line]		

```
Command: hide
Regenerating drawing.
Removing hidden lines: 25
```

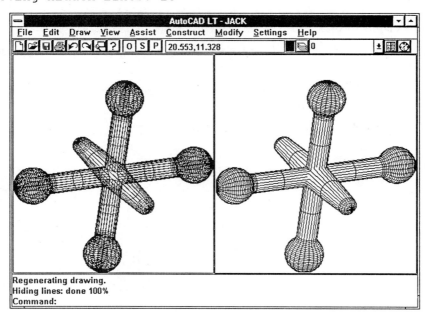

COMMAND OPTIONS
None.

RELATED AUTOCAD COMMANDS.
- **DView**   Removes hidden lines of perspective 3D views.
- **MView**   Removes hidden lines during plots of paper space drawings.
- **Plot**   Removes hidden lines during plotting of 3D drawings.
- **Regen**   Returns the view to wireframe.
- **Shade**   Performs quick renderings and quick hides of 3D models.
- **VPoint**   Select the 3D viewpoint.

Help *cont'd*

RELATED FILES
- **Aclt.Hlp**   The AutoCAD LT help file.
- **Ltutor.Hlp** A tutorial on using AutoCAD LT.
- **Whatsnew.Exe**
  Describes features new to AutoCAD LT Release 2.

TIPS
- Since **Help**, **?**, and **[F1]** are transparent commands, you can use them during another command to get help on the command's options.

- The text of the Windows help file is stored in the file Aclt.Hlp — this file is not ASCII and cannot be edited; it can be customized with the **Edit | Annotate** command.

Annotate   Add an annotation.
Bookmark   Mark the help topic with a bookmark; displays dialogue box:

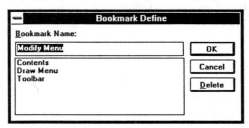

Help   Help on using the Windows help system.

BUTTON BAR OPTIONS

Contents   Displays the table of contents for the Help screens.
Search   Searches for keywords in the Help file.
Back   Moves back to the previous help screen.
History   Displays list of most-recently accessed help topics.
<<   Go back to previous topic.
>>   Access next help topic.
Glossary   Go to glossary page; displays dialogue box:

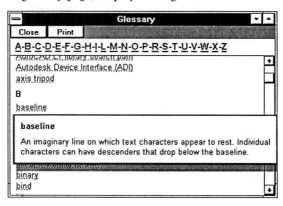

RELATED AUTOCAD COMMANDS
All.

RELATED SYSTEM VARIABLES
All.

 **'Help** *or* **'?** Rel. 1

Displays windows of information for using AutoCAD's commands.

Commands	Alt+	Menu bar	Function Key	Alias
'help	H,C	[Help]	[F1]	...
'?		[Contents]		

Command: **help**
Displays dialogue box:

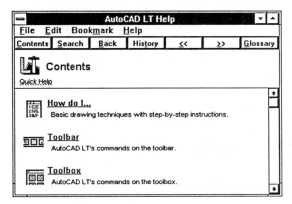

MENU BAR OPTIONS
File
   Open   Select HLP help file.
   Print Topic   Print current help topic.
   Print Setup   Setup printer.
   Exit   Exit Help.

Edit
   Copy   Copy help topic to the Windows Clipboard; displays dialogue box:

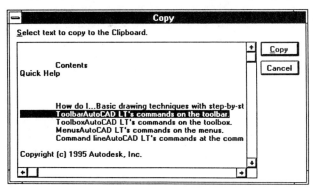

Advanced   Displays the **Advanced Options** dialogue box:

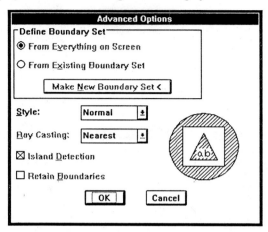

Style   Select hatching style: Normal, Outer, or Ignore.

RELATED AUTOCAD COMMANDS
- **BHatch**   Applies associative hatch pattern.
- **Hatch**   Applies non-associative hatch pattern.
- **Explode** Explodes a hatch pattern block into lines.

RELATED SYSTEM VARIABLES
- **HpAng**   Hatch pattern angle.
- **HpDouble** Doubled hatch pattern.
- **HpName**   Hatch pattern name.
- **HpScale** Hatch pattern scale.
- **HpSpace** Hatch pattern spacing.

HatchEdit

Rel. 2

Edits associative hatch patterns.

Command	Alt+	Menu bar	Alias	GC Alias
hatchedit	M,H	[Modify]	...	...
		[Edit Hatch]		

Command: `hatchedit`
Displays dialogue box:

COMMAND OPTIONS

Pattern Type   Toggle between:
      Predefined   Use hatch pattern found in Aclt.Pat.
      User-defined Create a hatch pattern on the fly.
      Custom   Use hatch pattern from another PAT file.

Pattern Properties:
  Pattern   Name of hatch pattern in Aclt.Pat or Acltiso.Pat file.
  Custom Pattern Name of custom hatch pattern.
  Scale   Hatch pattern scale.
  Angle   Global hatch pattern angle.
  Spacing   Spacing between pattern lines.
  Double   Double hatch (apply second pattern at 90 degrees).
  Exploded   Place hatch pattern as lines, rather than as a block.

Inherit Properties
      Select another hatch pattern to match parameters.
Associative   Toggle associativity of hatch pattern.
Advanced   Displays the **Advanced Options** dialogue box:

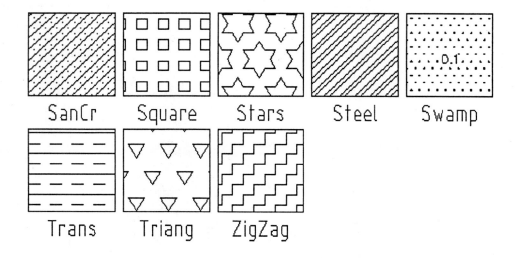

- Patterns are drawn at scale factor 1.0; those marked with a scale number (0.01, 0.5, etc) are shown at a smaller scale.

Hatch *cont'd*

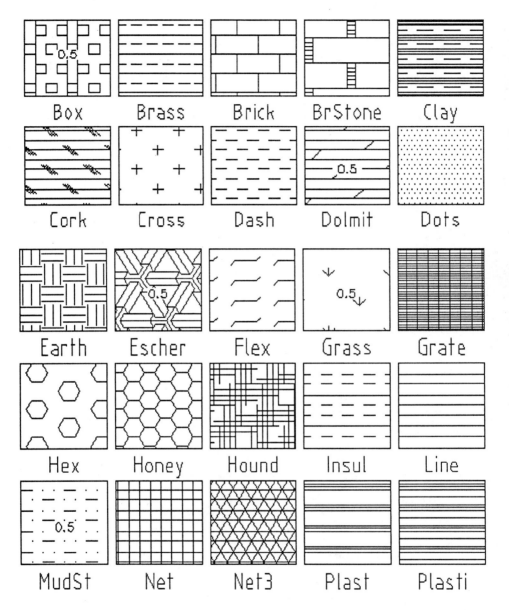

- Patterns are drawn at scale factor 1.0; those marked with a scale number (0.01, 0.5, etc) are shown at a smaller scale.

TIPS

- The **Hatch** command draws non-associative hatch patterns; the pattern remains in place when its boundary is edited.

- For complex hatch areas, you may find it easier to outline the area with a polyline (using the **Boundary** command) or to use the **BHatch** command.

- By default, the hatch is created as a block; the name begins with the letter "X" followed by a consecutive number. To create the hatch as line entities, precede the pattern name with * (asterisk).

- AutoCAD includes the following hatch patterns in the file Aclt.Pat:

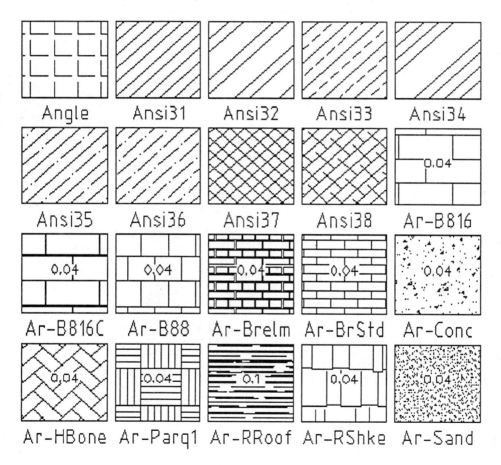

- Patterns are drawn at scale factor 1.0; those marked with a scale number (0.01, 0.5, etc) are shown at a smaller scale.

 **Hatch** Rel. 1

Draws a non-associative cross-hatch pattern within a closed boundary.

Command	Alt+	Menu bar	Alias	GC Alias
hatch	...	...	h	hs
				wh

```
Command: hatch
Pattern (? or name/U,style):
Scale for pattern <1.0000>:
Angle for pattern <0>:
Select objects: [pick]
Select objects: [Enter]
```

COMMAND OPTIONS

Pattern	Indicate name of hatch pattern.
?	List the hatch pattern names.
name	Specify hatch pattern name.
U	Create a user-defined hatch pattern.
style	Displays the sub-menu of hatching styles:
N	Hatch alternate boundaries (*short for Normal*).
O	Hatch only outermost boundaries (*short for Outermost*).
I	Hatch everything within boundary (*short for Ignore*).

RELATED AUTOCAD COMMANDS

- **BHatch** Automatic, associative hatching.
- **Boundary** Automatically creates polyline boundary.
- **Explode** Reduce hatch pattern to its constituent lines.
- **Snap** Change the hatch pattern's origin.

RELATED SYSTEM VARIABLES

- **HpAng** Hatch pattern angle.
- **HpDouble** Doubled hatch pattern.
- **HpName** Hatch pattern name.
- **HpScale** Hatch pattern scale.
- **HpSpace** Hatch pattern spacing.
- **SnapBase** Controls the origin of the hatch pattern.
- **SnapAng** Controls the angle of the hatch pattern.

RELATED FILES

- **Aclt.Pat** Hatch pattern definition file.
- **Acltiso.Pat** Hatch patterns for ISO (metric) drawings.

Handles   Rel. 1

Assigns a unique number to entities in the drawing; an undocumented command.

Command	Alt+	Menu bar	Alias	GC Alias
handles	...	...	...	...

```
Command: handles
Handles are enabled. Next handle: AD3
ON/DESTROY: destroy

* * * * *    W A R N I N G    * * * * *
Completing this command will destroy ALL
database handle information in the drawing.
Once destroyed, links into the drawing from
external database files cannot be made.

If you really want to destroy the database
handle information, please confirm this by
entering 'MAKE MY DATA' to proceed or 'NO'
to abort this command.

Proceed with handle destruction <NO>: NO
Command aborted. Database handles preserved.
```

COMMAND OPTIONS
DESTROY   Destroy all handles in the drawing.
ON   Turn handles on.

RELATED AUTOCAD COMMANDS
- DdModify   Displays the entity's handle number.
- List   Displays the entity's handle number.

RELATED SYSTEM VARIABLES
- Handles   Current state of handles (read only):
    0 On
    1 Off

TIPS
- AutoCAD preserves the handles of xref'ed drawings.

- AutoCAD displays one of six randomly selected messages:
    - DESTROY HANDLES
    - GO AHEAD
    - MAKE MY DATA
    - I AGREE
    - UNHANDLE THAT DATABASE
    - PRETTY PLEASE

Grid *cont'd*

RELATED SYSTEM VARIABLES
- **Gridmode**   Current grid visibility:
  - 0 Grid is off.
  - 1 Grid is on.
- **Gridunit**   Current grid x,y-spacing.

TIPS

■ The grid is most useful when set to the snap spacing or to a multiple of the snap spacing.

■ You can set a different grid spacing in each viewport and a different grid spacing in the x- and y-directions.

■ Rotate the grid with the **Snap** command's **Rotate** option.

■ The **Snap** command's **Isometric** option creates an isometric grid.

■ If a very dense grid spacing is selected, the grid will take a long time to display; press [Ctrl]+C to cancel the display.

■ AutoCAD does not display a too-dense grid and gives the message, "Grid too dense to display."

■ Grid markings are not plotted; to create a plotted grid, use the **Array** command to place an array of points.

'Grid

Rel. 1

Displays a grid of reference dots within the currently set limits.

Command	Alt+	Menu bar	Alias	GC Aliases
'grid	S,D,G	[Settings]	g	go
		[Drawing Aids]		gr
		[Grid]		gs

Command: **grid**
Grid spacing(X) or ON/OFF/Snap/Aspect <1.0000>:

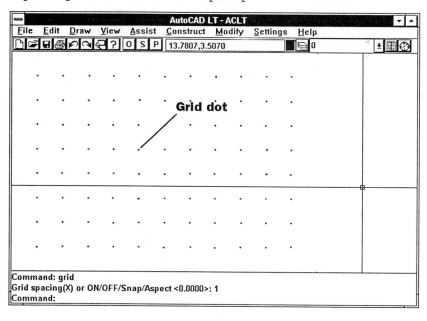

COMMAND OPTIONS
Aspect   Indicate different spacing for the x- and y-direction.
Grid spacing(X) Set the x- and y-direction spacing; an *X* following the value sets the grid spacing a multiple of the current snap setting.
OFF   Turn grid markings off.
ON   Turn grid markings on.
Snap   Make the grid spacing the same as the snap spacing.

RELATED AUTOCAD COMMANDS
- **Ddrmodes**   Sets the grid via a dialogue box.
- **Limits**   Sets the limits of the grid in WCS.
- **Snap**   Sets the snap spacing.

'GraphScr

Rel. 1

Switches the text screen back to the graphics screen in single-screen systems.

Command	Alt+	Function Key	Alias	GC Alias
'graphscr	...	[F2]	...	...

Command: **graphscr**
Displays the AutoCAD LT graphics window:

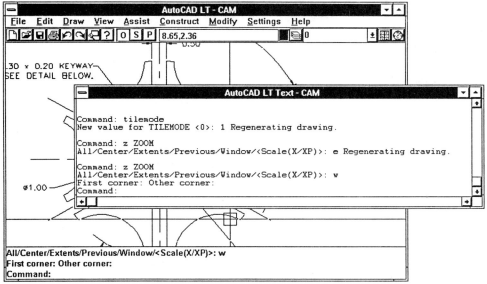

COMMAND OPTIONS
None.

RELATED AUTOCAD COMMANDS
- **Script**   Run script files, which can use the **GraphScr** command.
- **TextScr**   Switches from graphics screen to text screen.

RELATED SYSTEM VARIABLES
- **ScreenMode** Indicates whether current screen is text or graphics:
    - **0** Text screen.
    - **1** Graphics screen.
    - **2** Dual screen displaying both text and graphics.

TIP
- The **GraphScr** (and **TextScr**) commands do not work on a dual screen display.

'GetEnv

Rel. 2

Displays the value of variables in part of the Aclt.Ini initialization file.

Command	Alt+	Menu bar	Alias	GC Alias
'getenv	...	...	...	...

Command: **getenv**
Variable name:

Example output:
Command: **getenv**
Variable name: **monovectors**
1

COMMAND OPTIONS
All variables in [AutoCAD LT General] section of Aclt.Ini file..

RELATED AUTOCAD COMMANDS
- **Preferences** Dialogue box interface for setting some INI variables.
- **SetEnv** Sets the values of variables in the Aclt.Ini file.
- **SetVar** Reads and changes the values of system variables.

RELATED AUTOCAD FILES
- **Aclt.Ini** Contains the settings for AutoCAD's preferences and toolbox macros.

TIPS
- The **GetEnv** command only works with the **[AutoCAD LT General]** section of the Aclt.Ini file.

- If the variable does not exist, **GetEnv** returns 'nil', as follows:
    Command: **getenv**
    Variable name: **asdf**
    nil

- See Appendix B for the variables in the Aclt.Ini file.

- The **GetEnv** command does not exist in AutoCAD Release 12 or 13.

Fillet *cont'd*

- When the objects are on two different layers, the fillet is drawn on the current layer.
- The fillet radius must be smaller than the length of the lines. For example, if the lines to be filleted are 1.0m long, the fillet radius can be no more than 0.9999m.
- Use the C option of the **PLine** command to ensure a polyline is filleted at all vertices.
- The **Fillet** command cannot fillet polyline segments from different polylines.

Fillet

Rel. 1

Joins two intersecting lines, polylines, arcs, and circles with a radius.

Command	Alt+	Menu bar	Alias	GC Aliases
fillet	C,F	[Construct]	f	fi
		[Fillet]		fr

Command: **fillet**
Polyline/Radius/<Select first entity>: **[pick]**
Select second entity: **[pick]**

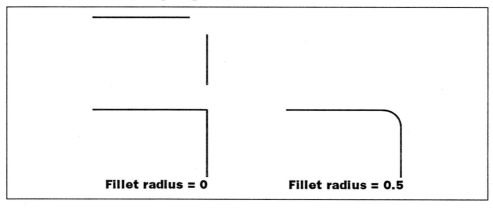

Fillet radius = 0   Fillet radius = 0.5

COMMAND OPTIONS
Polyline   Fillet all vertices of a polyline.
Radius   Indicate the filleting radius.

RELATED AUTOCAD COMMANDS
- Chamfer   Bevel intersecting lines or polyline vertices.
- Extend   Lengthen lines, polylines, and arcs.
- Trim   Shorten lines, polylines, and arcs.

RELATED SYSTEM VARIABLE
- FilletRad   The current filleting radius.

TIPS
- Pick the end of the entity you want filleted; the other end will remain untouched.

- The lines, arcs, or circles need not touch.

- Attempting to fillet two circles does not trim them.

- The **Fillet** command with a radius of 0.0 is a fast substitute for the **Extend** and **Trim** commands.

FIll *cont'd*

TIPS

- The state of fill (or no fill) does not come into effect until the next regeneration.

- Solids and polylines are not filled when the view is *not* in plan view, regardless of the setting of the **Fill** command.

- Since filled entities take longer to regenerate, redraw and plot, consider leaving fill off during editing and plotting.

- During plotting, use a wide pen for filled areas.

- **Fill** affects solids, traces, and all entities derived from polylines, including:
    - Donuts
    - Polygons.
    - Ellipses.
    - Variable-width polylines.

- AutoCAD LT cannot display filled and unfilled objects in different viewports.

'Fill

Rel. 1

Toggles wide entities (such as donuts, solids, and polylines) displayed and plotted as solid-filled or as outlines.

Command	Alt+	Menu bar	Alias	GC Alias
'fill	S,D,F	[Settings] [Drawing Aids] [Solid Fill]	fl	...

```
Command: fill
ON/OFF <On>:
```

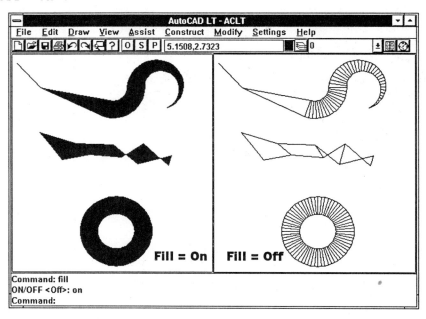

COMMAND OPTIONS
ON   Turn fill on, after next regeneration.
OFF   Turn fill off, after next regeneration.

RELATED SYSTEM VARIABLE
- Fillmode   Current setting of fill status:
    0 Fill mode is off.
    1 Fill mode is on (default).

RELATED AUTOCAD COMMAND
- Regen   Adjusts display to reflect fill-nofill status.

FileOpen

Rel. 2

Opens a DWG file without using a dialogue box.

Command	Alt+	Menu bar	Alias	GC Alias
fileopen	...	...	...	...

Command: **fileopen**
If drawing has not been saved, displays dialogue box:

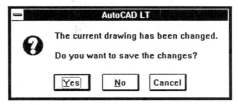

Enter name of drawing:

COMMAND OPTIONS
None.

RELATED AUTOCAD COMMAND
- **Open** Uses dialogue box to help you select the name of a drawing file.

RELATED SYSTEM VARIABLE
- **DbMod** Determines whether the drawing database has changed.

TIP
- The **FileOpen** command is used by AutoCAD during OLE operations.

- Filename must include the full subdirectory path name, unless the DWG file is in the current subdirectory.

TIPS
- These entities can be used as edges:
    - Line.
    - Polyline, including ellipse, donut, and polygon.
    - Arc.
    - Circle.
    - Viewport border.
    - Text.

- The trim location for a wide polyline is the center of the polyline.

- Pick the entity a second time to extend it to a second boundary line.

- Circles and other closed entities are valid edges: entity is extended in direction nearest to the pick point.

- Extending a variable-width polyline widens it proportionately; extending a splined poyline adds a vertex.

 # Extend

Rel. 2

Extends the length of a line, open polyline, or arc to a boundary.

Command	Alt+	Menu bar	Alias	GC Aliases
extend	M,X	[Modify]	ex	mx
		[Extend]		xt

```
Command: extend
Select boundary edge(s)...
Select entities: [pick]
Select entities: [Enter]
<Select entity to extend>/Undo: [pick]
<Select entity to extend>/Undo: [Enter]
```

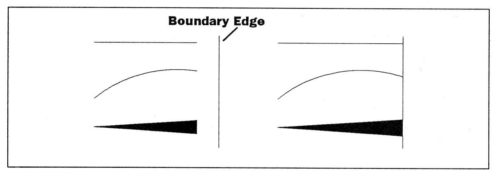

COMMAND OPTION
Undo   Undoes the most recent extend operation.

RELATED AUTOCAD COMMANDS
- **Change**   Changes the length of lines.
- **Stretch**   Stretches entities wider or narrower.
- **Trim**   Reduces the length of lines, polylines and arcs.

RELATED SYSTEM VARIABLES
None.

 # Explode

Rel. 1

Explodes a polyline, block, associative dimension, hatch, or associative hatch, into its constituent entities.

Command	Alt+	Menu bar	Aliases	GC Aliases
explode	M,L	[Modify]	ep	ce
		[Explode]	x	ex

```
Command: explode
Select entities: [pick]
```

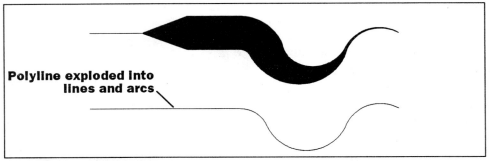

Polyline exploded into lines and arcs

COMMAND OPTIONS
None.

RELATED AUTOCAD COMMANDS
- **Block**   Recreates a block after an explode.
- **PEdit**   Converts a line into a polyline.
- **Undo**   Reverses the effects of explode.

TIPS
- For complex entities, such as blocks, you may have to use the **Explode** command more than once.

- The **Explode** command reduces:
    - **Block**   Constituent parts.
    - **Associative dimension**   Lines, solids, and text.
    - **Polyline**   Lines and arcs; width and tangency information is lost.

- You cannot explode xrefs, dependent blocks, or non-uniformly scaled blocks.

- The parts making up exploded blocks and associative dimensions may change color and linetypes.

- Resulting entities become the previous selection set.

 # Erase

Rel. 1

Erases entities from the drawing.

Command	Alt+	Menu bar	Alias	GC Aliases
erase	M,E	[Modify]	e	el
		[Erase]		er
				we

```
Command: erase
Select objects: [pick]
```

COMMAND OPTIONS
None.

RELATED AUTOCAD COMMANDS
- **Break**   Erases a portion of a line, circle, arc, or polyline.
- **Change** Changes the length of a line.
- **Oops**   Returns the most-recently erased entities to the drawing.
- **Trim**   Reduces the length of a line, polyline, or arc.
- **Undo**   Returns the erased entities to the drawing.

RELATED SYSTEM VARIABLES
- *None.*

TIPS
- The **Erase L** command combination erases the last-drawn item visible on the screen.

- The **Oops** command brings back the most-recently erased entities; use the **U** command to bring back other erased entities.

End

Rel. 1

Saves the drawing and exits AutoCAD to the operating system.

Command	Alt+	Menu bar	Alias	GC Alias
end	...	...	...	...

Command: **end**

COMMAND OPTIONS
None.

RELATED AUTOCAD COMMANDS
- SaveAs — Saves read-only drawings by another name.
- Quit — Leaves AutoCAD without saving the drawing.

RELATED SYSTEM VARIABLES
None.

TIPS
- AutoCAD renames the drawing file to .BAK before saving the contents of the drawing editor.

- The **End** command does not work with drawings set to read-only; use the **SaveAs** command instead.

Ellipse *cont'd*

Center, Endpoint, Rotation (ellipse construction method #4):

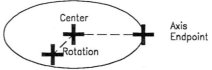

Center   Indicate the center point of the ellipse.
   Axis endpoint 2   Indicate the second endpoint of the major axis:
      Rotation   Indicate a rotation angle around the major axis.

Isometric circle (ellipse construction method #5):
Isocircle   Draws isometric circles; option appears only when **Snap** is set to isometric mode.

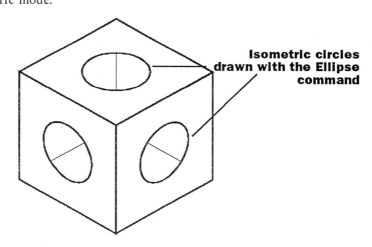

Isometric circles drawn with the Ellipse command

RELATED AUTOCAD COMMANDS
- **Isoplane**   Sets the current isometric plane.
- **PEdit**   Edits ellipses and other polyline entities.
- **Snap**   Controls the setting of isometric mode.

RELATED SYSTEM VARIABLES
- **SnapIsoPair** Current isometric plane:
    0 Left.
    1 Top.
    2 Right.
- **SnapStyl** Regular or isometric drawing mode:

TIPS
- Use ellipses to draw circles in isometric mode. When **Snap** is set to isometric mode, the **Ellipse** command's isocircle option projects a circle into the working isometric drawing plane.

- Use [Ctrl]+E to toggle isoplanes.

 # Ellipse

Rel. 1

Draws an ellipse by four different methods, as well as isometric circles.

Command	Alt+	Menu bar	Alias	GC Alias
ellipse	D,E	[Draw] [Ellipse]	el	ep

```
Command: ellipse
<Axis endpoint 1>/Center/Isocircle: [pick]
Axis endpoint 2: [pick]
<Other axis distance>/Rotation: [pick]
```

COMMAND OPTIONS

Endpoint, Endpoint, Axis (ellipse construction method #1):

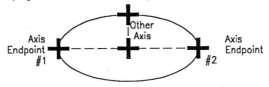

Axis endpoint 1 Indicate the first endpoint of the major axis:
  Axis endpoint 2 Indicate the second endpoint of the major axis:
    Other axis distance Indicate minor axis width.

Endpoint, Endpoint, Rotation (ellipse construction method #2):

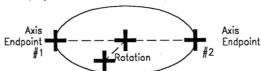

Axis endpoint 1 Indicate the first endpoint of the major axis:
  Axis endpoint 2 Indicate the second endpoint of the major axis:
    Rotation Indicate a rotation angle around the major axis.

Center, Endpoint, Axis (ellipse construction method #3):

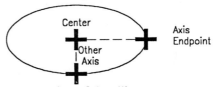

Center Indicate the center point of the ellipse.
  Axis endpoint 2 Indicate endpoint of major axis.
    Other axis distance Indicate minor axis width.

'Elev

Rel. 1

Sets elevation and thickness of extruded 3D entities (*short for ELEVation*).

Command	Alt+	Menu bar	Alias	GC Alias
'elev	S,E,E	[Settings] [Entity Modes]	ev	...

```
Command: elev
New current elevation <0.0000>:
New current thickness <0.0000>:
```

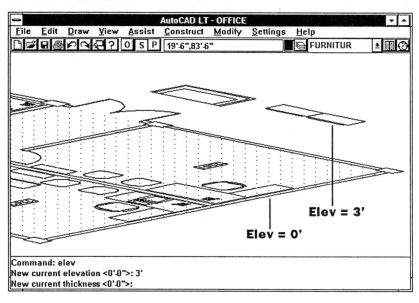

COMMAND OPTIONS
None.

RELATED AUTOCAD COMMANDS
- **Change** Changes the thickness and elevation of entities.
- **ChProp** Changes the thickness of entities.
- **Move**   Moves entities, including in the z-direction.

RELATED SYSTEM VARIABLES
- **Elevation** Stores the current elevation setting.
- **Thickness** Stores the current thickness setting.

TIPS
- The current value of elevation is used whenever a z-coordinate is not supplied.
- Thickness is measured up from the current elevation in the positive z-direction.

- Spell check the text, adding words to the custom dictionary as necessary.
- Save the file in ASCII or DOS Text format, keeping the .DXF extension.

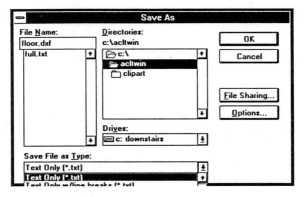

- Load the spell-checked file back into the drawing with the **DXFin** command.

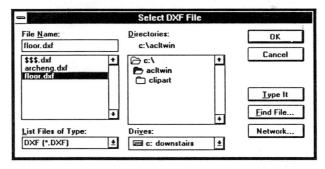

Spell check is complete!

QUICK START: Spell Check Drawing Text

The following work-around lets you ensure text in your drawings is spelled correctly:

- Use the **DXFout** command's **Entities** option and select all text objects. It may be helpful to freeze layers without text.

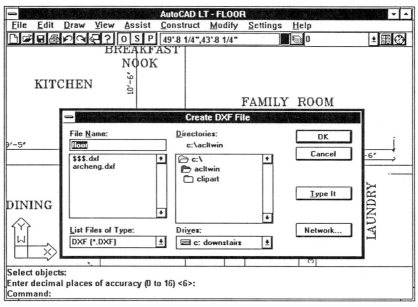

- Load DXF file into word processor with spell checker, such as WordPerfect or Word.

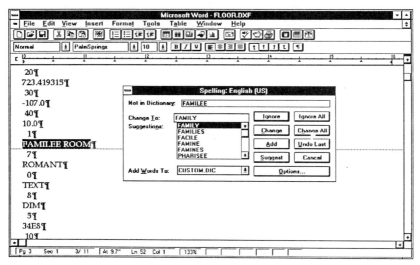

126 ■ The Illustrated AutoCAD LT for Windows Quick Reference

DxfOut

Rel. 1

Writes a DXF-format file of part or all of the current drawing (*short for Drawing interchange Format OUTput*).

Command	Alt+	Menu bar	Alias	GC Alias
dxfout	F,I,X	[File] [Import/Export] [DXF out]	dx	...

```
Command: dxfout
File name:
Enter decimal places of accuracy (0 to 16)/Entities/<6>: e
Select objects: [pick]
Select objects: [Enter]
Enter decimal places of accuracy (0 to 16) <6>:
```

COMMAND OPTIONS
0 — 16   Indicate decimal places of accuracy.
Entities   Select entities to export in DXF format.

RELATED AUTOCAD COMMANDS
- **DxfIn**   Reads a DXF-format file.
- **Save**   Writes the drawing in DWG format.

RELATED SYSTEM VARIABLES
- *None.*

TIPS
- Use the DXF format to exchange drawings with other CAD and graphics programs.

- Using the **Entities** option lets you import the DXF file into another AutoCAD drawing.

Dxfin

Rel. 1

Reads a DXF-format file into a drawing (*short for Drawing interchange Format INput*).

Command	Alt+	Menu bar	Alias	GC Alias
dxfin	F,I,D	[File] [Import/Export] [DXF in]	dn	...

Command: **dxfin**
File name:
Regnerating drawing.

COMMAND OPTIONS
None.

RELATED AUTOCAD COMMAND
- **DxfOut**   Writes a DXF-format file.

RELATED SYSTEM VARIABLES
- *None.*

TIPS
- When you attempt to import a full DXF file, AutoCAD complains: "Not a new drawing — only ENTITIES will be input."

- When AutoCAD detects an error in the DXF file, the **DXFin** command warns you — sometimes. If the error is in:
    - **Data**   The object appears missing from the drawing.
    - **Object name**   AutoCAD complains with a warning message, such as: "Unknown entity type LIasdfNE ignored on line 846" for a bad LINE.
    - **Table name**   AutoCAD complains with a warning message, such as: "Unknown table LAYasdfER on line 762" for a bad LAYER.

CLip   Displays the submenu for view clipping:
  Back   Displays the submenu for back clipping:
    ON   Turn on the back clipping plane.
    OFF   Turn off the back clipping plane.
    <Distance from target>
                    Indicate the location of the back clipping plane.
  Front   Displays the submenu for front clipping:
    Eye   Position the front clipping plane at the camera.
    <Distance from target>
                    Indicate the location of the front clipping plane.
  <Off>   Turn off view clipping.
Hide   Perform hidden-line removal.
Off   Turn off the perspective view.
Undo   Undo the most recent **DView** action.
<eXit>   Exit the **DView** command.

RELATED AUTOCAD COMMANDS
- **Hide**   Remove hidden-lines from a non-perspective view.
- **Pan**   Pan a non-perspective view.
- **VPoint**   Select an non-perspective viewpoint of a 3D drawing.
- **Zoom**   Zoom a non-perspective view.

RELATED SYSTEM VARIABLES
- **BackZ**   Back clipping plane offset.
- **FrontZ**   Front clipping plane offset.
- **LensLength** Perspective view lens length, in millimeters.
- **Target**   UCS 3D coordinates of target point.
- **ViewCtr**   2D coordinates of current view center.
- **ViewDir**   WCS 3D coordinates of camera offset from target.
- **ViewMode**   Perspective and clipping settings.
- **ViewSize**   Height of view.
- **ViewTwist**   Rotation angle of current view.

RELATED SYSTEM BLOCK
- **DViewBlock** Alternate viewing object during **DView**.

TIPS
- The view direction is from camera (your eye) to target.

- Press **[Enter]** at the 'Select objects:' prompt to display the house.

- You can replace the house block with your own custom symbol by redefining the DVIewBlock block.

- To view a 3D drawing in one-point perspective, use the **DView Zoom** command.

- Pull-down menus, transparent zoom and pan are not available during **DView**.

- Once the view is in perspective mode, you cannot use the **Zoom** and **Pan** commands.

DView

Rel. 1

Dynamically zooms and pans 3D drawings, and turns on perspective mode (*short for Dynamic VIEW*).

Command	Alt+	Menu bar	Alias	GC Alias
dview	V,Y	[View] [3D Dynamic View]	dv	...

Command: **dview**
Select objects: **[Enter]**
CAmera/TArget/Distance/POints/PAn/Zoom/TWist/CLip/Hide/Off/Undo/<eXit>:

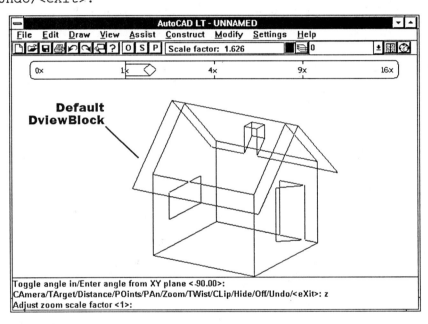

COMMAND OPTIONS

CAmera	Indicate the camera angle relative to the target:
Toggle	Switch between input angles.
TArget	Indicate the target angle relative to the camera.
Distance	Indicate the camera-to-target distance; turns on perspective mode.
POints	Indicate both the camera and target points.
PAn	Dynamically pan the view.
Zoom	Dynamically zoom the view.
TWist	Rotate the camera.

Style	Displays the style submenu:
Style name	Indicate a different style name.
?	List the currently loaded styles.
[Enter]	Exit the **DText** command.

COMMAND MODIFIERS

%%c	Draw diameter symbol: Ø.
%%d	Draw degree symbol: °.
%%o	Start and stop overlining.
%%p	Draw the plus-minus symbol: ±.
%%u	Start and stop underlining.
%%%	Draw the percent symbol: %.

RELATED AUTOCAD COMMANDS

- **DdEdit**   Edits the text.
- **Change**   Change the text height, rotation, style, and content.
- **Style**   Create new text styles.
- **Text**   Add new text to the drawing.

RELATED SYSTEM VARIABLES

- **TextSize**   The current height of text.
- **TextStyle**   The current style.
- **ShpName**   The default shape name

TIPS

- Use the **DText** command to easily place text in many locations on the drawing.

- Be careful: The spacing between lines of text does not neccesarily match the current snap spacing.

- The popdown menus are not available during the **DText** command.

- Transparent commands do not work during the **DText** command.

- You can enter any justification modes at the '<Start point>:' prompt.

- The command automatically repeats until cancelled with **[Enter]**.

 **DText** Rel. 1

Enters text in the drawing in a visual mode (*short for Dynamic TEXT*).

Command	Alt+	Menu bar	Aliases	GC Alias
dtext	D,T	[Draw]	t	tl
		[Text]	dt	

```
Command: dtext
Justify/Style/<Start point>: J
Align/Fit/Center/Middle/Right:
Height:
Rotation Angle:
Text:
```

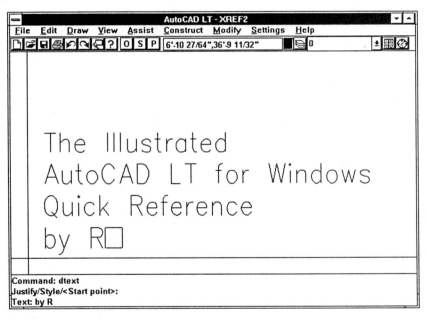

COMMAND OPTIONS
Justify　　　Displays the justification submenu:
　Align　　　　Align the text between two points with adjusted text height.
　Fit　　　　　Fit the text between two points with fixed text height.
　Center　　　Center the text along the baseline.
　Middle　　　Center the text horizontally and vertically.
　Right　　　　Right-justify the text.
<Start point>　Left-justify the text.

TempBev   Temporarily display the Aerial View window; disappears with the next command.

TOOLBAR OPTIONS

 Zoom to drawing extents.

 Zoom to drawing's virtual space.

 Zoom Aerial View to match current drawing view (the default).

 Toggle real-time pan and zoom.

 Zoom out by 2x.

 Zoom in by 0.5x.

 Bird's-eye view: display entire drawing in Aerial View (the default).

 Spyglass: display zoom-in view in Aerial View.

RELATED AUTOCAD COMMANDS
- **Pan**   Move the drawing view.
- **View**   Creates and displays named views.
- **Zoom**   Makes the view larger or smaller.

DsViewer

Rel. 1

Displays the bird's-eye view window; provides real-time pan and zoom (*short for "DiSplay Viewer"*).

Command	Alt+	Menu bar	Alias	GC Alias
dsviewer	S,A	[Settings] [Aerial View]	ds	...

Command: **dsviewer**
Displays Aerial View window:

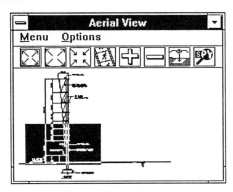

COMMAND OPTIONS

Menu:
   Set to Current Viewport
                Display current viewport.
   Exit   Turn off Aerial View.

Options:
   Big Icons   Display larger icons on toolbar.
   Tool Bar   Toggle display of the icon toolbar.
   Auto-Update Automatically update the display when editing changes occur.

UNDOCUMENTED OPTIONS

These Aerial View commands are not documented by Autodesk; they work by typing them at the 'Command:' prompt.

Cleanup   Clean up the display list, reducing memory consumption.
MemShow   List the amount of memory consumed by the display-list:

Donut or Doughnut

Rel. 1

Draws solid circles with a pair of wide polyline arcs.

Commands	Alt+	Menu bar	Alias	GC Alias
donut	D,D	[Draw] [Donut]	do	...
doughnut				

```
Command: donut
Inside diameter <0.5000>:
Outside diameter <1.0000>:
Center of doughnut: [pick]
Center of doughnut: [Enter]
```

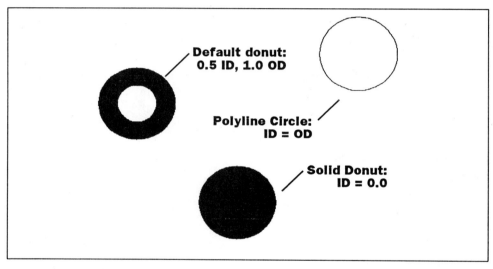

COMMAND OPTION
[Enter]   Exits the **Donut** command.

RELATED AUTOCAD COMMAND
Circle   Draws a circle.

RELATED SYSTEM VARIABLES
- DonutId   The current donut internal diameter.
- DonutOd   The current donut outside diameter.

TIP
- Command automatically repeats itself until cancelled by pressing **[Enter]**.

DLine *cont'd*

Start	Draw endcap only at startpoint.
Auto	Only draws endcaps when appropriate (default).

Dragline options:
```
Set dragline position to Left/Center/Right/<Offset from
  center = 0.0000>:
```
Left	Show dragline along left parallel line.
Center	Show dragline in center between both parallel lines.
Right	Show dragline along right parallel line.
Offset	Specify distance to offset the dragline from center of parallel lines (default = 0.000).

Snap options (causes DLine to start double line by snapping to an existing object):
```
Set snap size or snap On/Off.
```
Size	Prompts for new snap distance in pixels: 'New snap size (1 - 10) <3>:'.
OFF	Turns off object snap.
ON	Turns on object snap (default = 3 pixels).

Undo option:
Undo	Back up along previously drawn segments, erasing them one by one.

Width option (distance between parallel lines):
Width	Prompts for new width: 'New DLINE width <0.0500>:'.

CLose option:
CLose	Joins the current endpoint with the starting point.

Arc-specific options:
```
Break/CAps/CEnter/CLose/Dragline/Endpoint/Line/Snap/Undo/
Width/<second point>:
```
CEnter	Specify arc's center point.
Endpoint	Specify arc's endpoint.
Line	Switch back to parallel line drawing mode.

RELATED AUTOCAD COMMANDS
- **Line** — Draws single line segments.
- **PLine** — Draws connected line segments.

RELATED SYSTEM VARIABLE
- **LastPoint** — Coordinates of last drawn point.

# DLine

Rel. 1

Places two parallel lines with optional endcapping.

Command	Alt+	Menu bar	Alias	GC Aliases
dline	D,B	[Draw]	dl	me
		[Double Line]		db
				12

```
Command: dline
Break/Caps/Dragline/Offset/Snap/Undo/Width/<start point>:
```

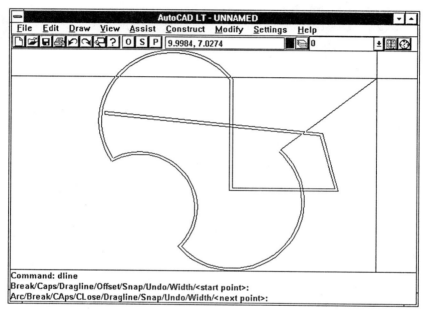

COMMAND OPTIONS

[Enter]   Continue drawing double line from point specified by LastPoint system variable.

Break options (determines if DLine cleans up intersections – creates breaks):
```
Break Dline's at start and end points?   OFf/<ON>:
```
  OF   Turn break option off.
  <ON>   Turn break option on (default); cleans up intersections.

Caps options:
```
Draw which endcaps?   Both/End/None/Start/<Auto>:
```
  Both   Draw a straight endcap at both ends of double line.
  End   Draw endcap only at endpoint.
  None   Draw no endcaps.

 # Divide

Rel. 2

Places points or blocks at an equally divided distance along lines, arcs, and polylines.

Command	Alt+	Menu bar	Alias	GC Alias
divide	C,E	[Construct] [Divide]	...	...

```
Command: divide
Select object to divide: [pick]
<Number of segments>/Block: B
Block name to insert:
Align block with object? <Y>:
Number of segments: 10
```

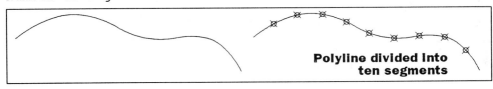

Polyline divided into ten segments

COMMAND OPTION
Block   Specify the name of the block to insert along the entity.

RELATED AUTOCAD COMMANDS
- Block   Create the block to use with the **Divide** command.
- Insert   Place a single block in the drawing.
- Measure   Divide an entity into measured distances.

RELATED SYSTEM VARIABLES
- PdMode   Sets the style of point drawn.
- PdSize   Sets the size of the point, in pixels.

TIPS
- Use **PdSize** and **PdMode** to make points visible along object.

- Minimum number of segments is 2; maximum is 32,767.

 **'Dist**

Rel. 1

Lists the 3D distance and angles between two points (*short for DISTance*).

Command	Alt+	Menu bar	Alias	GC Alias
'dist	A,D	[Assist] [Distance]	di	me

```
Command: dist
First point: [pick]
Second point: [pick]
```

Example result:
```
Distance=17.38, Angle in X-Y Plane=358, Angle from X-Y Plane=0
Delta X = 16.3000, Delta Y = -7.3000, Delta Z = 0.0000
```

COMMAND OPTIONS
None.

RELATED AUTOCAD COMMANDS
- Area   Calculates the area and perimeter of objects.
- Id   Lists the 3D coordinates of a point.
- List   Lists information about selected entities.

RELATED SYSTEM VARIABLE
- Distance Last calculated distance.

TIP
- Use object snaps to precisely measure between two geometric features. For example, use the CEN object snap to measure the distance between the centers of two circles.

Dim: Vertical

Rel. 1

Draws linear dimensions in the vertical direction.

Command	Alt+	Menu bar	Alias	GC Alias
vertical	D,I,V	[Draw] [Linear Dimensions] [Vertical]	d ver	1x

Dim: **vertical**
First extension line origin or RETURN to select: **[Enter]**
Select line, arc or circle: **[pick]**
Dimension line location(Text/Angle): **T**
Dimension text <>: **[Enter]**

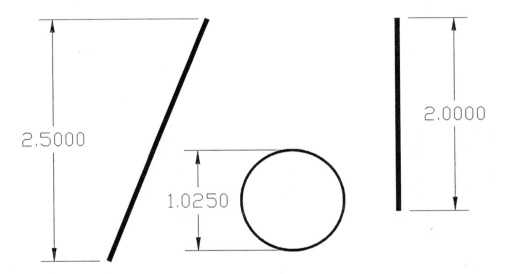

COMMAND OPTIONS
[Enter]   Displays the submenu for dimensioning objects:
  Select line, arc or circle
Text   Allows you to change the text of the dimension.
Angle   Specify the angle of the text.

RELATED DIM COMMAND
- **Aligned**   Draws linear dimensions aligned with objects.
- **Vertical**   Draws vertical dimensions.

Dim: Variables

Rel. 1

Similar to the **Dim:Status** command, but lists the dimensioning variables associated with a selected *associative* dimension.

Dim	Alt+	Menu bar	Alias	GC Alias
variables ...	...	...	d va	...

```
Dim: variables
Current dimension style: *UNNAMED
?/Enter dimension style name or RETURN to select
   dimension: [Enter]
Select dimension: [pick]
No style found.
Current dimension style: *UNNAMED
Status of *UNNAMED:
DIMALT    Off          Alternate units selected
DIMALTD   2            Alternate unit decimal places
DIMALTF   25.4000      Alternate unit scale factor

   ...

DIMTSZ    0.0000       Tick size
DIMTVP    0.0000       Text vertical position
DIMTXT    0.1800       Text height
DIMZIN    0            Zero suppression
```

COMMAND OPTIONS
None.

RELATED AUTOCAD COMMAND
- **DDim**   Changes dimvar settings via dialogue box.

RELATED SYSTEM VARIABLES
- *All*

Dim: Update   Rel. 1

Updates selected associative dimensions to the current units, text style, and dimension variables.

Dim	Alt+	Menu bar	Alias	GC Alias
update	M,I,U	[Modify]	d up	...
		[Edit Dimension]		
		[Update Dimension]		

```
Dim: update
Select objects: [pick]
Select objects: [Enter]
```

COMMAND OPTIONS
None.

RELATED DIM COMMANDS
- **Variables**   Controls appearance of dimensions.
- **Style**   Creates and sets text style for dimension.

RELATED SYSTEM VARIABLES
All dimension variables.

Dim: Undo

Rel. 1

Undoes the effect of the last dimension command.

Dim	Alt+	Menu bar	Alias	GC Alias
undo	. . .	. . .	d u	oo

`Dim: undo`

COMMAND OPTIONS
None.

RELATED DIM COMMANDS
All.

RELATED SYSTEM VARIABLES
None.

TIP
- Using the **Undo** (or **U**) command at the 'Command:' prompt undoes the entire dimensioning session.

Dim: TRotate

Rel. 1

Edits the location and orientation of text in associative dimensions.

Dim	Alt+	Menu bar	Alias	GC Alias
trotate	M,I,R	[Modify] [Edit Dimension] [Rotate Text]	d tr	...

```
Dim: trotate
Enter new text angle:
Select objects: [pick]
Select objects: [Enter]
```

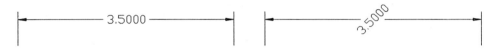

         Home   **Rotated Text**

COMMAND OPTION
Angle   Rotate the dimension text.

RELATED DIM COMMANDS
- **Hometext**   Returns dimension text to its original position.
- **TEdit**   Edits text in associative dimensions.

TIPS
- Use the **Rotate** command to rotate text in non-associative dimensions.
- An angle of 0 returns dimension text to its default orientations.

Dim: TEdit

Rel. 1

Edits the location and orientation of text in associative dimensions.

Dim	Alt+	Menu bar	Alias	GC Alias
tedit	M,I,M	[Modify] [Edit Dimension] [Move Text]	d te	...

```
Dim: tedit
Select dimension: [pick]
Enter text location (Left/Right/Home/Angle):
```

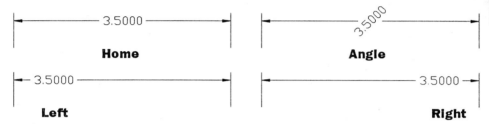

COMMAND OPTIONS

Angle	Rotate the dimension text.
Home	Return dimension text to original position.
Left	Move dimension text to the left.
Right	Move dimension text to the right.

RELATED DIM COMMANDS

- Edit — Edits associative dimension text.
- Style — Creates and sets associative dimension styles.

RELATED SYSTEM VARIABLES

- DimSho — Toggles whether dimension text dynamically updates while dragged.
- DimTih — Toggles whether text is drawn horizontal or aligned with dimension line.
- DimToh — Toggles whether dimension text is forces inside dimension lines.

TIPS

- Use the **DdEdit** command to edit text in non-associative dimensions.

- An angle of 0 returns dimension text to its default orientations.

- When entering dimension text with **DimTEdit's New** option, AutoCAD recognizes the following metacharacters:
  - [Square brackets]   Alternate units format string.
  - < Angle brackets >   Prefix and suffix text format string.

- Use the **Oblique** option to angle dimension lines by 30 degrees, suitable for isometric drawings. Use the **Style** command to oblique text by 30 degrees.

Dim: Style

Rel. 1

Selects a *text* style for dimension text; does *not* select a dimension style.

Dim	Alt+	Menu bar	Alias	GC Alias
style	...	...	d sty	...

Dim: **style**
New text style <STANDARD>:

COMMAND OPTION
New text style Type name of a text style defined in drawing.

INPUT OPTIONS
~dimvar   (Tilde prefix) Lists differences between current and selected dimstyle.
[Enter]   Lists dimvar settings for selected dimension object.

RELATED AUTOCAD COMMAND
■ Style   Creates new text styles.

RELATED SYSTEM VARIABLE
■ DimTxt   Height of dimension text.

TIP
■ At the **Dim:** prompt, the **Style** command sets the text style for dimension text and does *not* select a dimension style.

DimVar	Default	Meaning
DimTAD	Off	place Text Above the Dimension line
DimTFAC	1.0000	Tolerance text height scaling FACtor
DimTIH	On	Text Inside extensions is Horizontal
DimTIX	Off	place Text Inside eXtensions
DimTM	0.0000	Minus Tolerance
DimTOFL	Off	Force Line inside extension lines
DimTOH	On	Text Outside extensions is Horizontal
DimTOL	Off	generate dimension TOLerances
DimTP	0.0000	Plus Tolerance
DimTSZ	0.0000	Tick SiZe
DimTVP	0.0000	Text Vertical Position
DimTXT	0.1800	TeXT height
DimZIN	0	Zero suppression

COMMAND OPTIONS
None.

RELATED AUTOCAD COMMAND
- DDim   Changes dimvar settings via dialogue box.

RELATED SYSTEM VARIABLES
- *All*

AutoCAD LT R2 for Windows
KEYB...

MOUSE BUTTONS
① Pick
② Enter
③ Display cursor menu
Shift+② Cursor menu

FUNCTION KEYS
F1 Help
F2 Text-graphics screen
F5 Isoplane toggle
F6 Coordinate toggle
F7 Grid toggle
F8 Ortho toggle
F9 Snap toggle
F10 Menu bar

F3, F4, F11, and F12 are unused.

CONTROL KEYS
Ctrl+B snap toggle
Ctrl+C Cancel
Ctrl+D coorDinate toggle
Ctrl+E next isoplanE
Ctrl+G Grid toggle
Ctrl+O Ortho toggle
Ctrl+P menuecho toggle
Ctrl+S pauSe toggle
Ctrl+V next Viewport

COMMAND MODIFIERS
'   Transparent command:
    From point: **'zoom**
.   Force use of undefined command:
    Command: **.line**
_   Force English cmd in int'l version:
    Command: **_line**
Multiple Auto-repeat command:
    Command: **multiple circle**
'? Context-sensitive help:
    Command: **line '?**
~   Force display of dialogue box:
    Command: **insert ~**
tk Enter tracking mode:
    Command: **line**
    From point: **tk**

ALL select ALL objects
AU   AUtomatic: [pick] or BOX (default)
BOX Left to right: Crossing;
     Right to left: Window
C   Crossing
CP   Crossing Polygon
F   Fence
G   Group
L   Last
M   Multiple (no highlight)
P   Previous
SI   SIngle selection
U   Undo (remove from selection)
W   Window
WP   Window Polygon

Selection modes:
A   Add to selection set (default)
R   Remove from selection set

COLOR NUMBERS AND NAMES
BYLAYER   ...   Default color
0   ...   Background color
1   R   Red
2   Y   Yellow
3   G   Green
4   C   Cyan
5   B   Blue
6   M   Magenta
7   W   White
250 - 255 ...   Shades of grey
BYBLOCK   ...   Based on block

SPECIAL TEXT CHARACTERS
%%o   Overline.
%%u   Underline.
%%d   Degree symbol.
%%p   Plus-minus symbol.
%%%   Percent symbol.
%%nnn ASCII character *nnn*.

The Illustrated
AutoCAD LT® for Windows™
Quick Reference
Release 1 and 2

Ralph Grabowski

Press

I(T)P An International Thomson Publishing Company

Albany • Bonn • Boston • Cincinnati • Detroit • London • Madrid
Melbourne • Mexico City • New York • Pacific Grove • Paris • San Francisco
Singapore • Tokyo • Toronto • Washington

NOTICE TO THE READER

Publisher does not warrant or guarantee any of the products described herein or perform any independent analysis in connection with any of the product information contained herein. Publisher does not assume, and expressly disclaims, any obligation to obtain and include information other than that provided to it by the manufacturer.

The reader is expressly warned to consider and adopt all safety precautions that might be indicated by the activities described herein and to avoid all potential hazards. By following the instructions contained therein, the reader willingly assumes all risks in connection with such instructions.

The publisher makes no representations or warranties of any kind, including but not limited to, the warranties of fitness or particular purpose or merchantability, nor are any such representations implied with respect to the material set forth herein, and the publisher takes no responsibility with respect to such material. The publisher shall not be liable for any special, consequential or exemplary damages resulting, in whole or in part, from the readers' use of, or reliance upon, this material.

Book Design: Ralph H Grabowski, XYZ Publishing, Ltd. (ralphg@xyzpress.com)

Trademarks
AutoCAD LT is a registered trademark of Autodesk, Inc.
Windows is a trademark of the Microsoft Corporation.
All other product names are acknowledged as trademarks of their respective owners.

Copyright © 1995
by Delmar Publishers, Inc.
Autodesk Press imprint
an International Thomson Publishing Company.
The ITP logo is a trademark under license.

Printed in the United States of America

For information, contact:
Delmar Publishers, Inc.
3 Columbia Circle, Box 15015,
Albany, New York 12212-5015

All rights reserved. Certain portions of this work copyright 1993 - 1994. No part of this work covered by the copyright hereon may be reproduced or used in any form or by any means — graphic, electronic, or mechanical, including photocopying, recording, taping or information storage and retrieval systems — without written permission of the publisher.

1 2 3 4 5 6 7 8 9 10   xxx   00 99 98 97 96 95

Library of Congress Cataloging-in-Publication Data

Grabowski, Ralph.
    The illustrated AutoCAD LT for Windows quick reference : release 1 & 2 / Ralph Grabowski
        p. cm.
    Includes index.
    ISBN 0-8273-7828-9
    1. Computer graphics. 2. AutoCAD LT for Windows.   I Title.
T385.G6292   1995   95-21874
604.2'0285'5369--dc20   CIP

Contents

❷   *Command is new to AutoCAD LT Release 2.*
'   *(Apostrophe) Transparent command.*

The Layout of This Book xiv

A
'About 1
'Aperture 2
Arc 3
Area 6
Array 7
AttDef 10
'AttDisp 12
AttEdit 13
AttExt 15
    Quick Start: *Exporting Attributes from the Drawing*   **16**

B
'Base 18
❷ BHatch 19
'Blipmode 22
Block 23
BMake 25
❷ Boundary 27
Break 28

C
Chamfer 29
Change 30
ChProp 32
Circle 33
'Color *or* 'Colour 35
    Color Numbers, Names, and Abbreviations   **35**

Copy	36
CopyClip	38
CopyEmbed	39
CopyImage	40
CopyLink	41

D

DdAttDef	42
DdAttE	44
DdAttExt	45
DdChProp	46
DdEdit	48
'DdEModes	49
'DdGrips	52
DDim	54
DdInsert	57
'DdLModes	59
❷ DdModify	61
'DdOsnap	67
'DdPtype	68
DdRename	69
'DdRModes	70
'DdSelect	73
DdUcs	76
DdUcsP	78
'DdUnits	79
'Delay	81

DIM Commands

Dim	82
Dim1	84
Dim: Aligned	85
Dim: Angular	86
Dim: Baseline	87
Dim: Center	88

❷ *Command is new to AutoCAD LT Release 2.*   ' *(Apostrophe) Transparent command.*

Dim: Continue 89
Dim: Diameter 90
Dim: Exit 91
Dim: Hometext 92
Dim: Horizontal 93
Dim: Leader 94
Dim: Newtext 95
Dim: Oblique 96
Dim: Ordinate 97
Dim: Override 98
Dim: Radius 99
Dim: Redraw 100
Dim: Restore 101
Dim: Rotated 102
Dim: Save 103
Dim: Status 104
   Default values of dimension variables **104**
Dim: Style 106
Dim: TEdit 107
Dim: TRotate 108
Dim: Undo 109
Dim: Update 110
Dim: Variables 111
Dim: Vertical 112

'Dist 113
❷ Divide 114
DLine 115
Donut *or* Doughnut 117
DsViewer 118
   Aerial View toolbar icons **119**
DText 120
DView 122
DxfIn 124
DxfOut 125
   Quick Start: *Spell check drawing text* **126**

E

'Elev 128
Ellipse 129
End 131
Erase 132
Explode 133
❷ Extend 134

F

❷ FileOpen 136
'Fill 137
Fillet 139

G

❷ 'GetEnv 141
'GraphScr 142
'Grid 143

H

Handles 145
Hatch 146
    Hatch Pattern Library   **147**
❷ HatchEdit 150
'Help *or* '? 152
Hide 155

I

'Id 157
Insert 158
'Isoplane 161
    Quick Start: *Using Isometric Mode*   **162**

❷ *Command is new to AutoCAD LT Release 2.*   ' *(Apostrophe) Transparent command.*

L

'Layer 164
   Controlling Layers from the Toolbar **164**
'Limits 166
Line 168
'Linetype 169
   Linetype Library **170**
   Quick Start: *Creating a Custom Linetype* **171**
List 172
LogFileOff 173
LogFileOn 174
'LtScale 175

M

❷ Measure 176
Mirror 177
Move 178
MSlide 179
MSpace 180
Multiple 181
MView 182

N

New 184

O

Offset 190
Oops 191
Open 192
'Ortho 194
'OSnap 195
   Object Snap Icons **195**
   Object Snap Modes **196**

P

'Pan 198
'Paste Command 200
    Quick Start: Pasting Text In the Drawing   **201**
PasteClip 202
PEdit 204
Plan 207
PLine 208
Plot 210
Point 215
    Point Modes   **215**
Polygon 216
Preferences 218
PsOut 221
PSpace 224
    Quick Start: Enabling Paper Space   **225**
Purge 226

Q

QSave 227
QText 228
Quit 229

R

Rectang 230
Redo 231
'Redraw 232
Regen 233
Rename 234
'Resume 235
❷ RevDate 236
Rotate 238
'RScript 239

❷ *Command is new to AutoCAD LT Release 2.*   ' *(Apostrophe) Transparent command.*

S

Save	240
SaveAs	241
SaveDIB	242
Scale	244
'Script	245
Quick Start: Writing a Script File **246**	
Select	247
'SetEnv	248
'SetVar	249
Shade	250
Effect of ShadEdge **251**	
'Snap	252
Solid	253
Stretch	254
'Style	256
AutoCAD SHX Fonts **258**	
PostScript PFB Fonts **259**	

T

Text	260
Justification Modes **260**	
'TextScr	262
'Time	263
Toolbox	264
Icons Supplied with AutoCAD LT Release 2 **266**	
❷ Trim	267

U

U	268
Ucs	269
Definitions: User Coordinate System Terms **270**	
UcsIcon	271
Undo	272

'Units 273
'Unlock 275
Update 276
   Quick Start: Inserting an AutoCAD LT Object **276**

V
'View 279
VpLayer 280
VPoint 281
VPorts *or* ViewPorts 282
VSlide 285

W
WBlock 287
WmfIn 288
WmfOpts 290
WmfOut 291

X
XBind 292
Xref 293

Z
'Zoom 294

3
3dPoly 296

Programming LT
Toolbar Programming 297
   Special Characters **297**
     Quick Start: Changing a Toolbar/Toolbox Button **300**
     Quick Start: Writing a Macro **301**

                                                                   ' *(Apostrophe) Transparent command.*

Menu Programming 302
    Mouse Buttons **302**
    Understanding the Menu Structure **303**
    Quick Start: Creating a Simple Menu **305**
    Quick Start: Reloading a Menu File **306**
    Menu Macro Syntax **307**
    Quick Start: Advanced Menu Tricks **310**
    Quick Start: Designing an Icon Menu **312**
    Quick Start: Converting Slides into Icons **313**

Diesel Programming 315
    Quick Start: Diesel via ModeMacro **315**
    Diesel Functions **316**

Appendices

A: System Variables 321
    Dimension Variables **323**

B: Aclt.Ini Variables 334
    [AutoCAD LT General] Section **334**
    [AutoCAD LT Graphics Window] Section **337**
    [AutoCAD LT Text Window] Section **338**
    [AutoCAD LT ToolBox] Section **339**
    [WMF] Section **340**
    AcltDrv.Ini File **341**
    Win.Ini File **342**

C: Aliases and Icons 343
    Object Snap Modes **352**
    Command Modifiers and Point Filters **353**
    System Variables **353**

D: Removed Commands 354

Topical Index 359

The Layout of This Book

The *Illustrated AutoCAD LT for Windows Quick Reference* presents concise facts about all commands found in AutoCAD LT Release 1 and 2 for Windows. The clear format of this reference book demonstrates each command starting on its own page, plus these features:

- The commands that are undocumented or underdocumented by Autodesk.
- "Quick Start" mini-tutorials that help you get started quicker.
- Definitions of acronyms and hard-to-understand terms.
- Hundreds of context-sensitive tips.
- All system and initialization variables, including those not listed by the **SetVar** command, in Appendices A and B.
- A handy cross-reference of command names, AutoCAD and Generic CADD aliases, icon number and picture, in Appendix C.
- Commands missing from AutoCAD LT that are found in AutoCAD Release 12 and Release 13, in Appendix D.

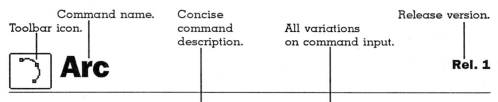

Draws a 2D arc of less than 360 degrees, by seven different methods.

Command	Alt+	Menu bar	Alias	GC Aliases
arc	D,A	[Draw]	a	a2
		[Arc]		a3

Command: **arc**
Center/<Start point>: [pick]
See below for the Arc command's seven arc-drawing options.

COMMAND OPTIONS
SCA (Start, Center, Angle) and SCE (Start, Center, End) arcs:

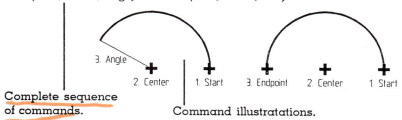

Command illustratations.

xiv ■ The Illustrated AutoCAD LT for Windows Quick Reference

COMMAND NAME & INPUT OPTIONS

The name of the command is in mixed upper and lower case, such as **DdAttExt**, to help explain the construction of the command name, which Autodesk tends to condense. Each command incudes all alternative methods of command input:

- Alternate spelling of command names, such as **Donut** and **Doughnut**.
- ' (the apostrophe prefix) indicates transparent commands, such as **'Blipmode**.
- AutoCAD alias, such as L for the **Line** command.
- Generic CADD alias, such as SL for the **Offset** command, as defined by the AcltGcad.Pgp file.
- Menu bar picks, such as **[Modify] [Stretch]** for the **Stretch** command.
- Control-key combinations, such as **[Ctrl]**+E for the **Isoplane** toggle.
- Function key, such as **[F1]** for the **Help** command.

The brief command description notes when the command is undocumented by Autodesk.

RELEASE NUMBERS

The release number indicates when the command first appeared in AutoCAD LT: **Rel. 1** or **Rel. 2**. This is useful when working with the two different versions.

RELATED COMMANDS & VARIABLES

Following the Command Options section, each command includes one or more of the following:

- Related AutoCAD commands.
- Related system variables.
- Related dimension variables.
- Related initialization variables.
- Related files.
- Related blocks.
- Input options.

DEFINITIONS & TIPS

Many commands include one or more tips that help you use the command more efficiently or warn you of the command's limitations. A number of commands include a list of definitions of acronyms and jargon words.

Ralph Grabowski
Abbotsford, British Columbia
September 20, 1995

Email: ralphg@xyzpress.com
or CompuServe 72700,3205

The Illustrated AutoCAD LT for Windows Quick Reference ∎ xv

For more information, contact:
Delmar Publishers
3 Columbia Circle , Box 15015
Albany, New York 12212-5015

International Thomson Publishing Europe
Berkshire House 168-173
High Holborn
London, WC1V7AA
England

Thomas Nelson Australia
102 Dodds Street
South Melbourne, 3205
Victoria, Australia

Nelson Canada
1120 Birchmont Road
Scarborough, Ontario
Canada M1K 5G4

International Thomson Editores
Campos Eliseos 385, Piso 7
Col Polanco
11560 Mexico D F Mexico

International Thomson Publishing GmbH
Königswinterer Strasse 418
53227 Bonn
Germany

International Thomson Publishing Asia
221 Henderson Road
#05 -10 Henderson Building
Singapore 0315

International Thomson Publishing - Japan
Hirakawacho Kyowa Building, 3F
2-2-1 Hirakawacho
Chiyoda-ku, Tokyo 102
Japan

'About

Rel. 1

Displays the AutoCAD LT release, serial number, licensee, and the Acad.Msg file.

Command	Alt+	Menu bar	Alias	GC Alias
'about	H,A	[Help] [About AutoCAD LT]	ab	...

Command: **about**
Displays dialogue box:

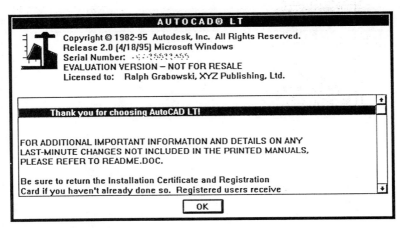

COMMAND OPTIONS
None.

RELATED AUTOCAD COMMANDS
None.

RELATED SYSTEM VARIABLE
- _PkSer   The AutoCAD software serial number.

RELATED FILE
- Aclt.Msg   The ASCII text file displayed by the **About** command.

TIP
- To change the message displayed by the **About** command, edit the \Acltwin**Aclt.Msg** file with a text editor, such as Notepad.

The Illustrated AutoCAD LT for Windows Quick Reference ■ **1**

'Aperture

Rel. 1

Sets the size, in pixels, of the object snap target height (or box cursor).

Command	Alt+	Menu bar	Alias	GC Alias
'aperture	A,O	[Assist] [Object Snap]	ap	to

Command: **aperture**
Object snap target height (1-50 pixels) <10>:

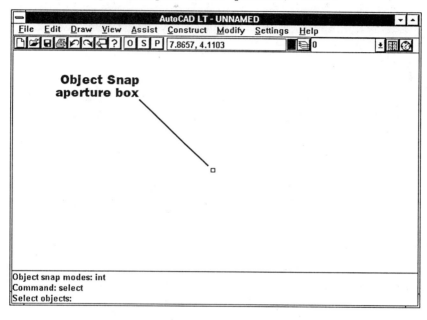

RELATED AUTOCAD COMMANDS
- **DdOSnap**   Sets the aperture size interactively.
- **DdSelect**   Sets the size of the object selection pickbox.
- **OSnap**   Sets the object snap modes.

RELATED SYSTEM VARIABLES
- **Aperture**   Contains the current target height:
    1 Minimum size.
    10 Default size.
    50 Maximum size.

TIP
- A larger aperture is easier to see; a smaller aperture "captures" fewer elements for selection.

 # Arc

Rel. 1

Draws a 2D arc of less than 360 degrees, by seven different methods.

Command	Alt+	Menu bar	Alias	GC Aliases
arc	D,A	[Draw]	a	a2
		[Arc]		a3

Command: **arc**
Center/<Start point>: **[pick]**
See below for the Arc command's seven arc-drawing options.

COMMAND OPTIONS
SCA (Start, Center, Angle) and SCE (Start, Center, End) arcs:

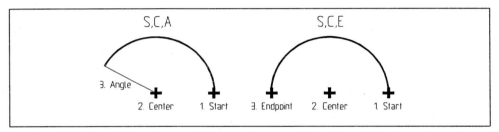

<Start point> Indicate the start point of a two-point arc:
  Center Indicate the center point of the arc:
    Included angle Indicate the arc's included angle; or
    <End point> Indicate the end point of the arc.

SEA (Start, End, Angle) arc:

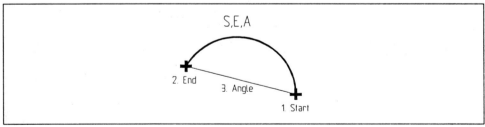

<Start point> Indicate the start point of a two-point arc:
  End Indicate the end point of the arc:
  Included angle Indicate the arc's included angle.

Arc *cont'd*

CSE (Center, Start, End) and CSA (Center, Start, Angle) arcs:

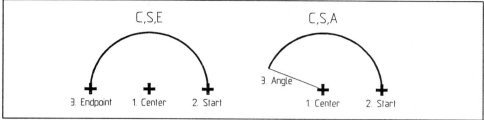

Center   Indicate the center point of a two-point arc:
  Start point   Indicate the arc's start point:
    <End point> Indicate the arc's end point; or
    Included angle Indicate the arc's included angle.

3P (three-point) arc:

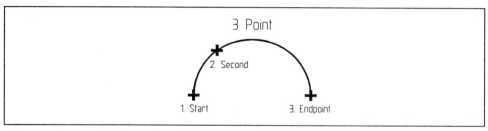

<Start point>   Indicate the arc's first point:
  <Second point>   Indicate the arc's second point:
    Endpoint   Indicate the arc's third point.

Continued arc:

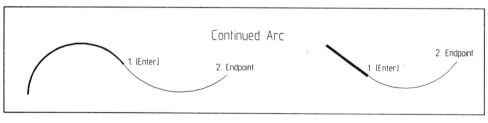

[Enter]   Press **[Enter]** at the '<Start point>:' prompt; AutoCAD continues arc from endpoint of last-drawn line or arc:
  Endpoint   Indicate arc's endpoint.

RELATED AUTOCAD COMMANDS
- **Circle**   Draws an arc of 360 degrees.
- **PLine**   Draws connected polyline arcs.
- **ViewRes**   Controls the roundness of arcs.

RELATED SYSTEM VARIABLE
- **LastAngle** Saves the included angle of the last-drawn arc (read-only).

TIPS
- To precisely start an arc from the endpoint of the last line or arc, press **[Enter]** at the 'Enter/<Start point>:' prompt.

- Angles are measured counterclockwise, unless specified clockwise with the **Units** command.

- You can only drag the arc during the last-entered option.

- Specifying an x,y,z-coordinate as the starting point of the arc draws the arc at the z-elevation.

- In some cases, it may be easier to draw a circle and use the **Break** and **Trim** commands to convert the circle into an arc.

- To draw an arc with width, use the **PLine** command's **Arc** option.

- AutoCAD LT does not support the length-of-chord option found in AutoCAD.

 **Area** Rel. 1

Calculates the area and perimeter of areas, closed entities, or polylines.

Command	Alt+	Menu bar	Alias	GC Alias
area	A,A	[Assist] [Area]	aa	...

```
Command: area
<First point>/Entity/Add/Subtract: [pick]
Next point: [pick]
Next point: [Enter]
Area = 1.8398, Perimeter = 6.5245
```

COMMAND OPTIONS
<First point> Indicate the first point to begin measurement.
Entity Indicate the entity to be measured.
Add Switch to add-area mode.
Subtract Switch to subtract-area mode.
[Enter] Indicate the end of the area outline.

RELATED AUTOCAD COMMAND
- **List** Lists all information of the selected entity.

RELATED SYSTEM VARIABLES
- **Area** Contains the most recently calculated area.
- **Perimeter** Contains the most recently calculated perimeter.

TIPS
- AutoCAD automatically "closes the polygon" before measuring the area.

- You can specify 2D x,y-coordinates or 3D x,y,z-coordinates.

- The **Object** option returns the following information:
    - **Circle, ellipse**   Area and circumference.
    - **Closed polyline, polygon**   Area and perimeter.
    - **Open objects**   Area and length.

- The area of a wide polyline is measured along its centerline; closed polylines must have only one closed area.

 # Array

Rel. 1

Creates a 2D rectangular or polar array of objects.

Command	Alt+	Menu bar	Alias	GC Alias
array	C,A	[Construct] [Array]	ar	rc

Command: **array**
Select objects: **[pick]**
Select objects: **[Enter]**

Rectangular array:
Rectangular or Polar array (R/P): **r**
Number of rows (—) <1>:
Number of columns (| | |) <1>:
Unit cell or distance between rows (—):
Distance between columns (| | |):

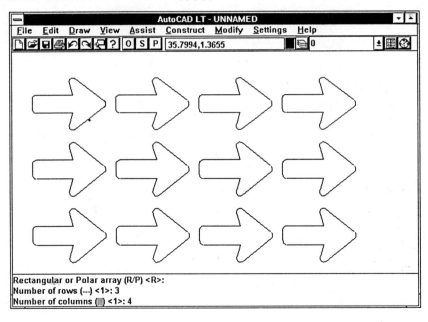

The Illustrated AutoCAD LT for Windows Quick Reference ■ 7

Array *cont'd*

Rotated Polar array:
```
Rectangular or Polar array (R/P): p
Center point of array: [pick]
Number of items:
Angle to fill (+=ccw, -=cw) <360>:
Rotate objects as they are copied? <Y> [Enter]
```

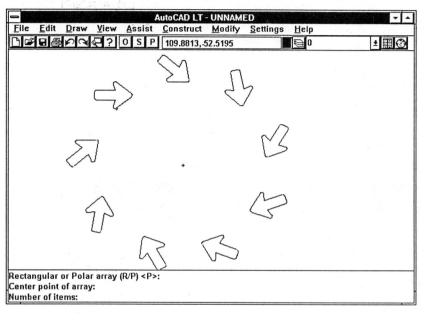

Non-rotated Polar array:
```
Rectangular or Polar array (R/P): p
Center point of array: [pick]
Number of items:
Angle to fill (+=ccw, -=cw) <360>:
Rotate objects as they are copied? <Y> N
```

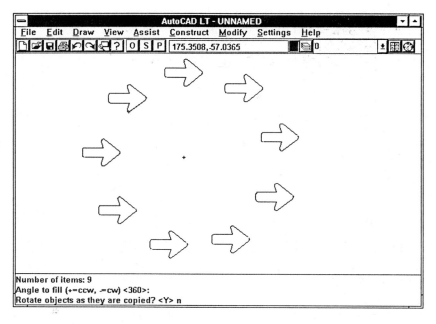

COMMAND OPTIONS
R   Create a rectangular array of the selected object.
P   Create a polar array of the selected object.

RELATED AUTOCAD COMMAND
- Copy   Creates one or more copies of the selected object.

RELATED SYSTEM VARIABLE
- SnapAng   Determines the angle of rectangular arrays.

TIPS
- To array at an angle, use the **Rotation** option of the **Snap** command.

- Rectangular array draws:
    - **Up**   In the positive x-direction.
    - **Right** In the positive y-direction
    - **Down**   Negative row distance.
    - **Left**   Negative column distance.

- Polar arrays are drawn in the counterclockwise direction; to draw the array in the opposite direction, specify a negative angle.

- Maximum size of array is 32,767 items.

AttDef

Rel. 1

Defines attribute modes and prompts (*short for ATTribute DEFinition*).

Command	Alt+	Menu bar	Alias	GC Aliases
attdef	...	...	ad	ac
				as
				at

```
Command: attdef
Attribute modes — Invisible:N Constant:N Verify:N Preset:N
Enter (ICVP) to change, RETURN when done: [Enter]
Attribute tag:
Attribute prompt:
Default attribute value:
Justify/Style/<Start point>: [pick]
```

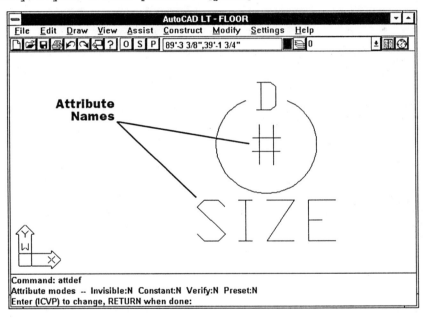

COMMAND OPTIONS

Attribute mode Select the modes for the attribute:
- **I** — Toggles visibility of attribute text in drawing (*short for Invisible*).
- **C** — Toggles fixed or variable value of attribute (*short for Constant*).
- **V** — Toggles confirmation prompt during input (*short for Verify*).
- **P** — Toggles automatic insertion of default values (*short for Preset*).

Justify Select the justification mode for the attribute text.
Style Select the text style for the attribute text.
<Start point> Indicate the start point of the attribute text.

RELATED AUTOCAD COMMANDS
- **AttDisp**   Controls the visibility of attributes.
- **AttEdit**   Edits the values of attributes.
- **AttExt**   Extracts attributes to disk.
- **AttRedef**   Redefines an attribute or block.
- **Block**   Binds attributes to objects.
- **DdAttDef**   Define attributes via a dialogue box.
- **DdAttE**   Extract the values of attributes via a dialogue box.
- **DdEdit**   Edits the values of attributes via a dialogue box.
- **Insert**   Inserts a block and prompts for attribute values.

RELATED SYSTEM VARIABLES
- **AFlags**   Contains the value of modes in bit form:
    - 0 No attribute mode selected.
    - 1 Invisible.
    - 2 Constant.
    - 4 Verify.
    - 8 Preset.
- **AttDia**   Toggles use of dialogue box during **Insert** command:
    - 0 Uses command-line prompts.
    - 1 Uses dialogue box.
- **AttReq**   Toggles use of defaults or user prompts during **Insert** command:
    - 0 Assume default values of all attributes.
    - 1 Prompts for attributes.

TIPS
- Constant attributes cannot be edited.

- Attribute *tags* cannot be null (have no value); attribute *values* may be null.

- When you press **[Enter]** at the '<Start point>:' prompt, **AttDef** automatically places the next attribute below the previous one.

'AttDisp

Rel. 1

Controls the display of all attributes in the drawing (*short for ATTribute DISPlay*).

Command	Alt+	Menu bar	Alias	GC Alias
'attdisp	...	...	at	ad

```
Command: attdisp
Normal/ON/OFF <Normal>:
```

Normal: Computer

ON: Computer Touch Pentium 60 $3,105

OFF:

COMMAND OPTIONS

Normal   Display attributes according to Attdef setting.
ON   Display all attributes, regardless of **AttDef** setting.
OFF   Display no attribute, regardless of **AttDef** setting.

RELATED AUTOCAD COMMAND
- **AttDef**   Defines new attributes, including their default visibility.

RELATED SYSTEM VARIABLE
- **AttMode**   Contains current setting of **AttDisp**:
  - 0 Off.
  - 1 Normal.
  - 2 On.

TIPS
- If **RegenAuto** is off, use the **Regen** command after **AttDisp** to see changes to attributes.

- If you define invisible attributes, **AttDisp** lets you turn them on.

AttEdit

Rel. 1

Edits attributes in a drawing (*short for ATTribute EDIT*).

Command	Alt+	Menu bar	Alias	GC Alias
attedit	...	...	ae	...

```
Command: attedit
Edit attributes one at a time? <Y> [Enter]
Block name specification <*>:
Attribute tag specification <*>:
Attribute value specification <*>:
Select Attributes: [pick]
1 attributes selected.
Value/Position/Height/Angle/Style/Layer/Color/Next <N>:
```

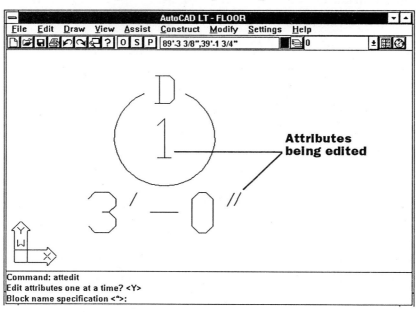

COMMAND OPTIONS

Value	Change or replace the value of the attribute:
Change	Change some of the attribute value.
Replace	Replace attribute with a new value.
Position	Move the text insertion point of the attribute.
Height	Change the attribute text height.
Angle	Change the attribute text angle.
Style	Change the text style of the attribute text.
Layer	Move the attribute to a different layer.

The Illustrated AutoCAD LT for Windows Quick Reference ■ 13

AttEdit *cont'd*

Color	Change the color of the attribute text.
Next	Edit the next attribute.

RELATED AUTOCAD COMMANDS
- **AttDef**   Defines an attribute's original value and parameter.
- **AttDisp**   Toggles an attribute's visibility.
- **AttEdit**   Edits the values of attributes.
- **DdAttDef**   Define attributes via a dialogue box.
- **DdAttE**   Edits the values of an attribute.
- **DdEdit**   Edits the values of an attribute.
- **Explode**   Reduces an attribute to its tag.

TIPS
- Constant attributes cannot be edited with **AttEdit**.

- You can only edit attributes parallel to the current UCS.

- Unlike other text input to AutoCAD, attribute values are case-sensitive.

- To edit null attribute values, use **AttEdit**'s global edit option and enter \ (backslash) at the 'Attribute value specification:' prompt.

- The wildcard characters ? and * are interpreted literally at the 'String to change:' and 'New String:' prompts.

AttExt

Rel. 1

Extracts attribute data from the drawing to a file on disk (*short for ATTribute EXTract*).

Command	Alt+	Menu bar	Alias	GC Alias
attext	...	...	ae	...

Command: **attext**
CDF, SDF or DXF Attribute extract (or Entities)? <C>:
Displays the Select Template File and Create Extract File dialogue boxes.

COMMAND OPTIONS
CDF   Output attributes in comma-delimited format.
SDF   Output attributes in space-delimited format.
DXF   Output attributes in DXF format.
Entities   Select entities to extract attributes from.

RELATED AUTOCAD COMMANDS
- AttDef   Defines attributes.
- DdAttExt   Define attribute extraction via a dialogue box.

RELATED FILES
- *.Txt   Required extension for the template file.
- *.Txt   Extension for CDF and SDF extraction files.
- *.Dxx   Extension for DXF extraction files.

TIPS
- To output the attributes to the printer, specify:
  CON   to appear on the text screen.
  PRN or LPT1   to print on parallel port 1.
  LPT2 or LPT3   to print to parallel 2 or 3.

- Before you can specify the SDF or CDF option, you must create the template file.

- CDF files use the following conventions:
  - Specified field widths are the maximum width.
  - Positive number fields have a leading blank.
  - Character fields are enclosed in ' ' (single quote marks).
  - Trailing blanks are deleted.
  - Null strings are '' (two single quote marks).
  - Uses spaces; do not uses tabs.
  - Use the C:DELIM and C:QUOTE records to change the field and string delimiters to another character.

AttExt *cont'd*

QUICK START: Exporting Attributes

How to extract attribute data from a drawing:

1. CREATE TEMPLATE FILE

If you want the attribute extracted in CDF or SDF format, you must first create a template file; the DXF format does not use a template file. The *template file* is used by the **AttExt** (or **DdAttExt**) command to: (1) determine which attributes to extract; and (2) the format of the extracted information. The template file uses the following format codes:

- The **Type** describes the type of attribute data:
    - **C**   Alpha-numeric characters.
    - **N**   Numbers only.

- The **Width** (www) describes the field width from 001 to 999 characters wide, padded with leading zeros.

- The **Precision** (ddd) describes the number of decimal places from 001 to 999 (i.e., 0.1 to 0.00000...001), padded with leading zeros.

Field Name	Type,Width,Precision	Description
BL:NAME	Cwww000	Name of block
BL:NUMBER	Nwww000	Number of occurances
CHAR_ATTRIBUTE_TAG	Cwww000	Character attribute tag
NUMERIC_ATTR_TAG	Nwwwddd	Numeric attribute tag
BL:LAYER	Cwww000	Block's layer name
BL:ORIENT	Nwwwddd	Block rotation angle
BL:LEVEL	Nwww000	Block's nesting level
BL:X	Nwwwddd	Block insertion x-coordinate
BL:Y	Nwwwddd	Block insertion y-coordinate
BL:Z	Nwwwddd	Block insertion z-coordinate
BL:XSCALE	Nwwwddd	Block's x-scale factor
BL:YSCALE	Nwwwddd	Block's y-scale factor
BL:ZSCALE	Nwwwddd	Block's z-scale factor
BL:XEXTRUDE	Nwwwddd	Block's x-extrusion
BL:YEXTRUDE	Nwwwddd	Block's y-extrusion
BL:ZEXTRUDE	Nwwwddd	Block's z-extrusion
BL:HANDLE	Cwwwddd	Block's handle hex-number.

Shell out of AutoCAD and use a text editor to create a template file.

Here is an example:

Field	Template	Description
BL:NAME	C008000	*8-character block name*
BL:NUMER	N004000	*Number of occurrences*
VENDOR	C016000	*Vendor attribute (16 chars)*
MODELNO	N012000	*Model # attr (12 digits)*

Save the file as ASCII text with the .TXT extension and return to AutoCAD.

2. SELECT OUTPUT FORMAT
Use either the **AttExt** or **DdAttExt** command to extract attribute data. Decide on the output format:
- **CDF**   Comma-delimited format (the default), best for importing into a spreadsheet; sample output:
  ```
  'Desk',55,'Steelcase',2248599597
  ```
- **DXF**   Drawing interchange format, similar to an entities-only DXF file.
- **SDF**   Space-delimited format, best for importing into a database program; sample output:
  ```
  Desk       55           Steelcase    2248599597
  ```

3. SELECT OBJECTS
Either select the blocks you want to extract attributes from, or select all objects in the drawing. AutoCAD ignores all non-block objects and blocks with no attributes.

4. SPECIFY TEMPLATE FILE
Enter the name of the template file you created earlier.

5. SPECIFY OUTPUT FILENAME (Optional)
Specify any filename — *except* the template file's name. If you do not specify an output filename, AutoCAD uses the drawing's name, appending:
- **TXT**   to CDF and SDF files.
- **DXX**   to DXF files.

6. CLICK [OK]
Click the OK button and AutoCAD places extracted attribute data into the output file. AutoCAD will stop if it finds any errors in the format of the template file, or if the selection set contains no attributes.

'Base

Rel. 1

Changes the 2D or 3D insertion point of a drawing; located at (0,0,0) by default.

Command	Alt+	Menu bar	Alias	GC Alias
'base	S,W,B	[Settings] [Drawing] [Base]	ba	bp

Command: **base**
Base point <0.0000,0.0000,0.0000>:

COMMAND OPTIONS
None.

RELATED AUTOCAD COMMANDS
- Block   Allows you to specify the insertion point of a new block.
- Insert   Inserts another drawing into the current drawing.
- Xref   References another drawing.

RELATED SYSTEM VARIABLE
- InsBase   Contains the current setting of the drawing insertion point.

TIPS
- When you attempt to give the z-coordinate a value, AutoCAD LT warns: "Beware. Z insertion base is not zero."

- The **Base** command is only needed when you want to insert the drawing into another drawing.

 BHatch

Rel. 2

Automatically applies an associative hatch pattern within a boundary (*short for Boundary HATCH*).

Command	Alt+	Menu bar	Alias	GC Alias
bhatch	D,H	[Draw] [Boundary Hatch]	...	...

Command: **bhatch**
Displays dialogue box:

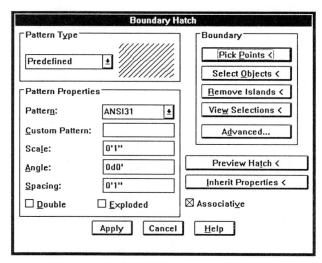

COMMAND OPTIONS

Pattern Type Select type of hatch pattern.
Pattern Properties:
  Pattern Name of hatch pattern (defined in Acad.Pat).
  Custom Pattern Specify name of a user-defined pattern.
  Scale Hatch pattern scale (default: 1.0).
  Angle Hatch pattern angle (default: 0 degrees).
  Spacing Spacing between lines of a user-defined hatch pattern.
  Double Toggles double hatching.
  Exploded Places hatch pattern as lines, rather than as a block.

Preview Hatch Preview the hatch pattern before it is applied.
Inherit Properties
      Select the hatch pattern parameters from an existing hatch pattern.

BHatch *cont'd*

Boundary:
    Pick Points   Pick points that define the hatch pattern boundary.
    Select Objects   Select objects to be hatched.
    Remove Islands   Remove islands from the hatch pattern selection set.
    View Selections   View hatch pattern selection set.
    Advanced...   Displays the **Advanced Options** dialogue box:

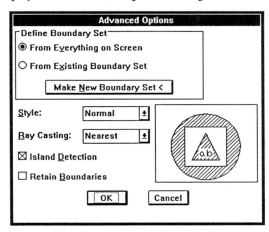

Define Boundary Set:
    From Everything on Screen   All objects visible in current view.
    From Existing Boundary Set   Current boundary selection set.
    Make New Boundary Set   Dialogue boxes disappear to allow you to select the boundary.

Style   Hatching style:
    Normal   Alternate areas are hatched.
    Outer   Only the outermost areas are hatched.
    Ignore   Everything within boundary is hatch.

Ray Casting   Determines how AutoCAD searches for the hatch boundary:
    Nearest   From pick point to nearest object.
    +X   From pick point in +x direction.
    -X   From pick point in -x direction.
    +Y   From pick point in +y direction.
    -Y   From pick point in -y direction.

Retain Boundaries
    Toggles whether the boundary is retained after hatch is placed.

RELATED AUTOCAD COMMANDS
- **Boundary**   Automatically traces a polyline around a closed boundary.
- **Explode**   Reduces a group of hatch patterns.
- **Hatch**   Creates a non-associative hatch within a manually selected perimeter.

RELATED SYSTEM VARIABLES
- **HpAng**   Current hatch pattern angle (default = 0).
- **HpDouble**   Single or double hatching:
    0 Single (default).
    1 Double.
- **HpName**   Current hatch pattern name (up to 31 characters long):
    ""   No current hatch pattern name.
    "."   Eliminate current name.
    "ANSI31"   Default name.
- **HpScale**   Current hatch pattern scale factor (default = 1.0000).
- **HpSpace**   Current hatch pattern spacing factor (default = 1.0000).
- **SnapBase**   Starting x,y-coordinates of hatch pattern (default = 1.0000, 1.0000).

RELATED FILE
- **Aclt.Pat**   Hatch pattern definition file.

TIPS
- See the **Hatch** command for a list of the hatch patterns supplied with AutoCAD LT.

- The **BHatch** command first generates a boundary polyline, then hatches the inside area. Use the **Boundary** command to obtain just the bounding polyline.

- The associative hatch pattern created by AutoCAD Relase 13 cannot be read by AutoCAD LT.

- **BHatch** stores hatching parameters in the pattern's extended entity data; use the **List** command to display parameters stored in extended entity data, as shown below by Hatch pattern, Hatch scale, and Hatch angle:

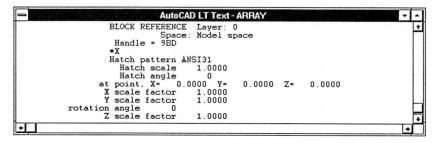

'Blipmode Rel. 1

Turns the display of pick-point markers, known as "blips," off and on.

Command	Alt+	Menu bar	Alias	GC Alias
'blipmode	S,D,B	[Settings] [Drawing Aids] [Blips]	bm	pc

Command: **blipmode**
ON/OFF <On>: off

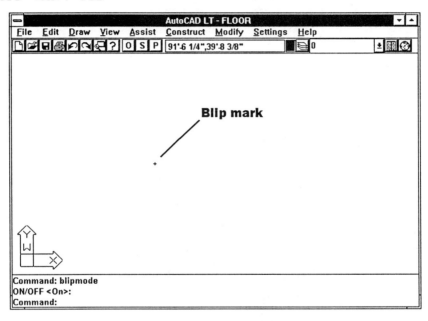

COMMAND OPTIONS
ON Turn on display of pick point markers.
OFF Turn off display of pick point markers.

RELATED AUTOCAD COMMANDS
- **DdRModes** Allows blipmode toggling via a dialogue box.
- **Redraw** Cleans blips off the screen.

RELATED SYSTEM VARIABLE
- **Blipmode** Contains the current setting of blipmode.

TIP
- You cannot change the size of the blipmark.

Block

Rel. 1

Defines a group of objects as a single named object; creates symbols.

Command	Alt+	Menu bar	Alias	GC Aliases
block	...	...	b	pc
				cc
				cn
				cw

```
Command: block
Block name (or ?):
Insertion base point: [pick]
Select objects:
```

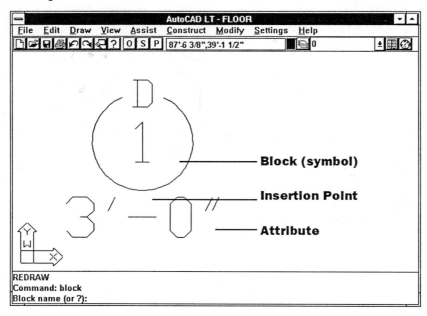

COMMAND OPTIONS
?   Lists the blocks stored in the drawing.

RELATED AUTOCAD COMMANDS
- **Explode**   Reduces a block into its original objects.
- **Insert**   Adds a block or another drawing to the current drawing.
- **Oops**   Returns objects to the screen after creating the block.
- **WBlock**   Writes a block to a file on disk as a drawing.
- **XRef**   Displays another drawing in the current drawing.

Block *cont'd*

RELATED FILES
- *.DWG   All drawing files are insertable as blocks.

TIPS
- A block name has up to 31 alpha-numeric characters, including the dollar ($), dash (-), and underscore (_) characters, such as "A_$700_gold-plated_screwdriver". However, blocks stored on disk (using the **WBlock** command) are limited to eight characters.

- AutoCAD works with five types of blocks:
    - **User block**   Block created by you.
    - **Nested block**   A block within a block.
    - **External reference** Externally referenced drawings.
    - **Dependent block**   Blocks in an externally referenced drawing.
    - **Unnamed block**   Blocks created by certain AutoCAD command, such as hatch patterns created by the **Hatch** command.

- Some of the unnamed blocks created by AutoCAD:
    - *Dn Associative dimension block.
    - *Un Unnamed block created by **BMake** command.
    - *WMFn WMF (picture) block; first block is *WMF0.
    - *Xn Hatch pattern block; first block is *X1.

- Use the **BMake** command's **List Block Names** option to see the complete list of unnamed blocks in the drawing.

- Use the **INSertion** object snap to select the block's insertion point.

- A block created on:
    - Layer 0 is inserted on the current layer.
    - Any other layer is always inserted on that same layer.

BMake

Rel. 1

Creates a block (symbol) via dialogue box.

Command	Alt+	Menu bar	Alias	GC Alias
bmake	C,B	[Construct] [Make Block]	...	...

Command: **bmake**
Displays dialogue box:

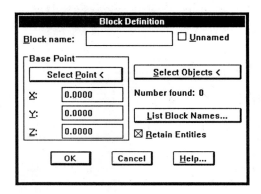

COMMAND OPTIONS

Block name	Name of the block.
Unnamed	Create a block with the generic name of *U*n*.
Base point	Indicate the insertion point of the block:
Select point	Pick a point on the screen
X	Specify the x-coordinate.
Y	Specify the y-coordinate.
Z	Specify the z-coordinate.
Select objects	Select the objects that make up the block.
Number found	Number of objects making up the block.
Retain Entities	Do not erase the objects after making the block.

BMake *cont'd*

List Block Names
: Displays a dialogue box listing named and unnamed blocks found in the drawing:

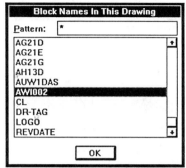

Pattern   Use the wildcard characters * and ? to shorten the list of block names.

RELATED AUTOCAD COMMANDS
- **Block**   The command-line version of the **BMake** command.
- **DdInsert**   Insert a block using a dialogue box.
- **Oops**   Returns objects erased during creation of the block.

TIPS
■ The **BMake** command's **List Block Names** option fails when the drawing contains too many blocks.

■ The **BMake** command is unique to AutoCAD LT and does not occur in AutoCAD Release 12 or 13.

 **Boundary** Rel. 2

Creates a boundary as a polyline.

Command	Alt+	Menu bar	Alias	GC Alias
boundary	...	[Draw] [Boundary]	...	...

Command: **boundary**
Displays dialogue box:

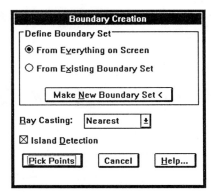

COMMAND OPTIONS
Define Boundary Set:
  From everything on screen   All objects visible in current view.
  From existing boundary set   Current boundary selection set.
  Make new boundary set   Dialogue box disappears so you can select boundary.

Ray Casting   Determines how AutoCAD searches for the hatch boundary; from your pick point to the:
  Nearest   Nearest object.
  +X   +x direction.
  -X   -x direction.
  +Y   +y direction.
  -Y   -y direction.

Island Detection
    Toggles whether islands are detected.

RELATED AUTOCAD COMMANDS
- **PLine**   Draws a polyline.
- **PEdit**   Edits a polyline, such as the boundary.

TIPS
- Use the **Boundary** command together with the **Offset** command to create poching.
- Although the **Boundary Creation** dialogue box looks identical to the **BHatch** command's **Advanced Options** dialogue box, be aware there are differences.

Break

Rel. 1

Removes a portion of a line, polyline, arc, or a circle.

Command	Alt+	Menu bar	Alias	GC Alias
break	M,B	[Modify] [Break]	br	ob

```
Command: break
Select object: [pick]
Enter second point (or F for first point): f
Enter first point: [pick]
Enter second point: [pick]
```

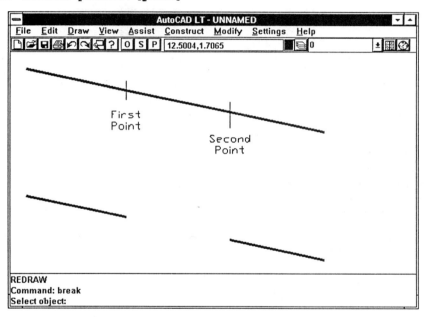

COMMAND OPTIONS
F   Specify the first break point.
@   Use the first break point coordinates for the second break point.

RELATED AUTOCAD COMMANDS
- **Change**   Changes the length lines.
- **PEdit**   Removes and relocates vertices of polylines.
- **Trim**   Shortens the lengths of lines.

Chamfer

Rel. 1

Bevels the intersection of two lines, or all vertices of a 2D polyline.

Command	Alt+	Menu bar	Alias	GC Aliases
chamfer	C,H	[Construct]	cf	ca
		[Chamfer]		ch

```
Command: chamfer
Polyline/Distances/<Select first line>: D
Enter first chamfer distance:
Enter second chamfer distance:
Polyline/Distances/<Select first line>: [pick]
Select second line: [pick]
```

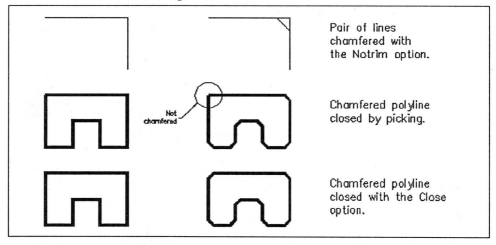

Pair of lines chamfered with the Notrim option.

Chamfered polyline closed by picking.

Chamfered polyline closed with the Close option.

COMMAND OPTIONS
Polyline Chamfer all vertices of a polyline.
Distances Specify the chamfer distances.

RELATED AUTOCAD COMMAND
- Fillet   Rounds the intersection with a radius.

RELATED SYSTEM VARIABLES
- ChamferA   First chamfer distance.
- ChamferB   Second chamfer distance.

Change

Rel. 1

Modifies the color, elevation, layer, linetype, linetype scale, and thickness of any object, and certain properties of lines, circles, blocks, text, and attributes.

Command	Alt+	Menu bar	Alias	GC Aliases
change	...	...	cf	dg
		...		mp

```
Command: change
Select objects: [pick]
Select objects: [Enter]
Properties/<Change point>: p
Change what property (Color/Elev/LAyer/LType/Thickness)?
```

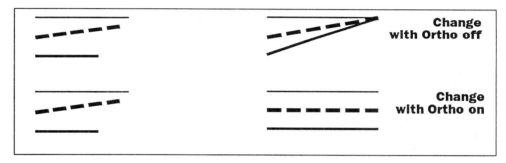

COMMAND OPTIONS

<Change point> Pick an entity to change:
- **[pick line]** Indicate the new length of line.
- **[pick circle]** Indicate the new radius of circle.
- **[pick block]** Indicate the new insertion point or rotation angle of a block.
- **[pick text]** Indicate the new location of text.
- **[pick block]** Indicate an attribute's new text insertion point, text style, new height, rotation angle, text, tag, prompt, or default value.

[Enter]   Change the insertion point, style, height, rotation angle, and string of text.

Properties   Change properties of the object, as follows:
- **Color**   Change the color of the entity.
- **Elev**   Change the elevation of the entity.
- **LAyer**   Move the entity to a different layer.
- **LType**   Change the linetype of the entity.
- **Thickness**   Change the thickness of any entity, except blocks.

RELATED AUTOCAD COMMANDS
- **ChProp**   Contains the properties portion of the **Change** command.
- **Color**   Changes the current color setting.
- **DdChProp**   Dialogue box for changing entity properties.
- **DdModify**   Changes most aspects of all objects.
- **Elev**   Changes the working elevation and thickness.

RELATED SYSTEM VARIABLES
- **CeColor**   The current color setting.
- **CeLType**   The current linetype setting.
- **CircleRad** The current circle radius.
- **CLayer**   The name of the current layer.
- **Elevation** The current elevation setting.
- **TextSize**   The current height of text.
- **TextStyle** The current text style.
- **Thickness** The current thickness setting.

TIPS
- The **Change** command cannot change:
    - Size of donuts.
    - Radius and length of arcs.
    - Length of polylines.
    - Justification of text.

- Use the **Change** command to change the endpoints of a group of lines to a common vertex.

- Turn ortho mode on to extend or trim a group of lines, without needing a cutting edge (as the **Extend** and **Trim** commands).

- The **DdModify** command is more powerful than the **Change, ChProp,** and **DdChProp** commands but works on just one entity at a time.

ChProp

Rel. 1

Modifies the color, layer, linetype, and thickness of most entities.

Command	Alt+	Menu bar	Alias	GC Aliases
chprop	...	...	cr	cg
				wg

```
Command: chprop
Select objects: [pick]
Select objects: [Enter]
Change what property (Color/LAyer/LType/Thickness)?
```

COMMAND OPTIONS
Color   Change the color of the object.
LAyer   Move the object to a different layer.
LType   Change the linetype of the object.
Thickness   Change the thickness of any object except blocks.

RELATED AUTOCAD COMMANDS
- Change   Also allows changes to lines, circles, blocks, text, and attributes.
- Color   Changes the current color setting.
- DdChProp   Dialogue box version of the **ChProp** command.
- DdModify   Changes most aspects of all objects.
- Elev   Changes the working elevation.

RELATED SYSTEM VARIABLES
- CeColor   The current color setting.
- CeLtype   The current linetype name.
- CLayer   The name of the current layer.
- Thickness   The current thickness setting.

TIP
Use the **Change** command to change the elevation of an object.

 # Circle

Rel. 1

Draws 2D circles by four different methods.

Command	Alt+	Menu bar	Alias	GC Aliases
circle	D,C	[Draw]	cr	c2
		[Circle]		c3

```
Command: circle
3P/TTR/<Center point>: [pick]
Diameter/<Radius>: [pick]
```

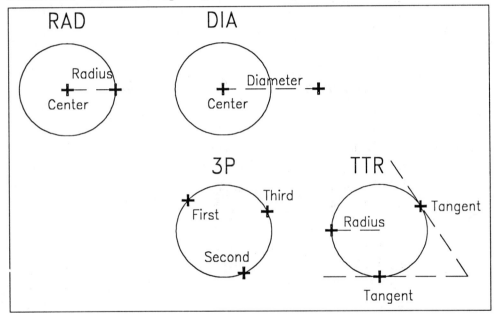

COMMAND OPTIONS

Center, Radius circle and Center, Diameter circle:
<Center point> Indicate the circle's center point:
   <Radius>   Indicate the circle's radius.
   Diameter   Indicate the circle's diameter.

Three-point circle:
3P   Draw a three-point circle:
First point: Indicate first point on circle; and
   Second point: Indicate second point on circle; and
   Third point: Indicate third point on circle.

Circle *cont'd*

Tangent, Tangent, Radius circle:
TTR   Draw a circle tangent to two lines:
  Enter Tangent spec:
                Indicate first point of tangency; and
  Enter second Tangent spec:
                Indicate second point of tangency; and
  Radius: Indicate first point of radius; and
  Second point: Indicate second point of radius.

RELATED AUTOCAD COMMANDS
- **Arc**   Draws an arc.
- **Donut**   Draws a solid-filled circle or donut.
- **Ellipse** Draws an elliptical circle.
- **ViewRes** Controls the visual roundness of circles.

RELATED SYSTEM VARIABLE
- **CircleRad**   The current circle radius.

TIPS
- Sometimes it is easier to create an arc by drawing a circle, then using the **Break** or **Trim** commands to convert the circle into an arc.

- Giving a circle thickness turns it into a cylinder.

'Color or 'Colour

Rel. 1

Sets the new working color.

Command	Alt+	Menu bar	Alias	GC Aliases
'color	S,E,C	[Settings]	...	cs
'colour		[Entity Modes]		lc
		[Colors]		

```
Command: color
New entity color <BYLAYER>:
```

COMMAND OPTIONS

BYLAYER   Set working color to color of current layer.
BYBLOCK   Set working color of inserted blocks.
Color Number   Set working color using number (1 to 255), name, or abbreviation:

Color name	Number	Abbreviation
Red	1	R
Yellow	2	Y
Green	3	G
Cyan	4	C
Blue	5	B
Magenta	6	M
White	7	W
Greys	250 - 255	...

RELATED AUTOCAD COMMANDS
- Change   Changes the color of entities.
- ChProp   Changes the color of entities.
- DdEModes   Sets new working color using a dialogue box.
- DdChProp   Changes the color of entities via a dialogue box.

RELATED SYSTEM VARIABLE
- CeColor   The current entity color setting:
             1 (red)   minimum value
             7 (white) default value
             255 (grey) maximum value

TIPS
- 'BYLAYER' means that objects take on the color assigned to that layer.
- 'BYBLOCK' means that objects take the color in effect when the block is inserted.
- White entities display as black when the background color is white.
- The **Preferences** command's **Color** option changes all objects to monochrome.

 # Copy

Rel. 1

Creates one or more copies of an object.

Command	Alt+	Menu bar	Alias	GC Aliases
copy	C, C	[Construct]	cp	co
		[Copy]		oc
				wc

```
Command: copy
Select objects: [pick]
Select objects: [Enter]
<Base point or displacement>/Multiple: [pick]
Second point of displacement: [pick]
```

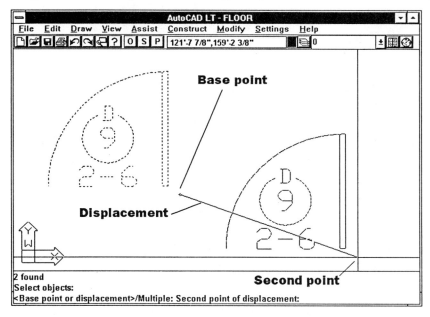

COMMAND OPTIONS

<Base point or displacement>
        Indicate the starting point, or the distance to move.
Second point of displacement
        Indicate the point to move.
Multiple   Allows an object to be copied more than once.
[Ctrl]+C   Cancels multiple object copying.

RELATED AUTOCAD COMMANDS
- **Array**   Draws a rectangular or polar array of objects.
- **Move**   Moves an object to a new location.
- **Offset**   Draws parallel lines, polylines, circles and arcs.

TIPS
- Use the M (multiple) option to quickly place several copies of the original object.

- Inserting a block multiple times is more efficient than placing multiple copies.

- Turn ortho mode on to copy objects in a precise horizontal and vertical direction.

- Turn snap mode on to copy objects in precise increments.

- Use object snap modes to precisely copy objects from one geometric feature to another.

- To copy an object by a known distance, enter 0,0 as the 'Base point.' Then, enter the known distance as the 'Second point.'

- **Copy** works in 2D (supply coordinate pairs) and 3D (supply coordinate triplets). In 2D, the current elevation is used as the z-coordinate.

 # CopyClip

Rel. 1

Copies selected objects to the Windows Clipboard (*short for COPY to CLIPboard*).

Command	Alt+	Menu bar	Alias	GC Alias
copyclip	E,V	[Edit] [Copy Vectors]	cc	...

Command: **copyclip**
Select objects: **[pick]**
Select objects: **[Enter]**

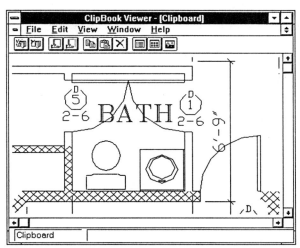

COMMAND OPTIONS
None.

RELATED AUTOCAD COMMANDS
- **CopyEmbed** Copy to the Clipboard in OLE format.
- **CopyImage** Copy windows area to the Clipboard in raster format.
- **CopyLink** Copy current viewport to the Clipboard.
- **SaveDib** Export selected objects in the current view to a BMP file.
- **WmfOut** Export selected obejcts to a WMF file.

TIPS
- AutoCAD's **CopyClip** command sends the selected objects to the Windows Clipboard in the following formats:
  - **Native** AutoCAD R13 drawing format.
  - **Picture** WMF (Windows metafile) vector format.

- When specifying the All option to the 'Select objects:' prompt, the **CopyClip** command only selects all objects visible in the current viewport.

CopyEmbed

Rel. 1

Copies selected objects to the Windows Clipboard for OLE embedding.

Command	Alt+	Menu bar	Alias	GC Alias
copyembed	E,E	[Edit] [Copy Embed]	ce	...

```
Command: copyembed
Select objects: [pick]
Select objects: [Enter]
```

COMMAND OPTIONS
None.

RELATED AUTOCAD COMMANDS
- **CopyClip**   Copy selected objects to the Windows Clipboard.
- **CopyLink**   Copy objects to the Clipboard for OLE linking.
- **CopyImage** Copy windows area to the Clipboard in raster format.
- **SaveDIB**   Export selected objects in the current view to a BMP file.
- **Update**   Update the embedded drawing.
- **WmfOut**   Export selected obejcts to a WMF file.

TIPS
- AutoCAD LT's **CopyEmbed** command sends the selected objects to the Windows Clipboard in the following formats:
    - **Native**   AutoCAD DWG Release 12 format.
    - **OwnerLink**   Name of object's source application and filename.
    - **Picture**   WMF (Windows metafile) vector format.
- In the other application, use the **Edit | Paste** or **Paste Special** commands to paste the AutoCAD image into the document; the **Paste Special** command lets you specify the pasted format:

CopyImage

Rel. 1

Copies a rectangular area to the Windows Clipboard in raster format.

Command	Alt+	Menu bar	Alias	GC Alias
copyimage	E,I	[Edit] [Copy Image]	ci	...

Command: **copyimage**
Select an area of the screen. [pick] [pick]

COMMAND OPTIONS
None.

RELATED AUTOCAD COMMANDS
- **CopyClip** Copy selected objects to the Windows Clipboard.
- **CopyEmbed** Copy objects to the Clipboard for OLE embedding.
- **CopyLink** Copy objects to the Clipboard for OLE linking.
- **SaveDIB**   Export part of AutoCAD window to a BMP file.
- **WmfOut**   Export selected objects to a WMF file.

RELATED WINDOWS COMMANDS
- **[Prt Scr]**   Copy the entire screen to the Windows Clipboard.
- **[Alt]+[Prt Scr]** Copy the topmost window (such as a dialogue box) to the Clipboard.

TIPS
- AutoCAD LT's **CopyImage** command sends the selected area to the Windows Clipboard in just one format:
    - **Bitmap** BMP (bitmap) raster format.

- In the other Windows application, use the **Edit | Paste** command to paste the AutoCAD image into the document.

 # CopyLink — Rel. 1

Copies the current viewport to the Windows Clipboard for OLE linking.

Command	Alt+	Menu bar	Alias	GC Alias
copylink	E,L	[Edit] [Copy Link]	cl	...

Command: `copylink`

COMMAND OPTIONS
None.

RELATED AUTOCAD COMMANDS
- **CopyClip** Copy selected objects to the Windows Clipboard.
- **CopyEmbed** Copy objects to the Clipboard for OLE embedding.
- **CopyImage** Copy windows area to the Clipboard in raster format.
- **SaveDIB**   Export selected objects in the current view to a BMP file.
- **Update**   Updates linked objects.
- **WmfOut**   Export selected objects to a WMF file.

TIPS
- AutoCAD LT's **CopyClip** command sends the selected objects to the Windows Clipboard in the following formats:
  - **Bitmap**   BMP (bitmap) raster format.
  - **ObjectLink** Embedding information for OLE objects.
  - **OwnerLink** Name of object's source application and filename.
  - **Picture**   WMF (Windows metafile) vector format.

- In the other application, use the **Edit | Paste** or **Edit | Paste Special** commands to paste the AutoCAD image into the document; the **Paste Special** command lets you specify the pasted format.

- In the other Windows application, use the **Edit | Insert Object** command to select AutoCAD LT to insert a drawing image into the document:

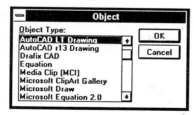

DdAttDef

Rel. 1

Define an attribute definition via a dialogue box (*short for Dynamic Dialogue ATTribute DEFinition*).

Command	Alt+	Menu bar	Alias	GC Alias
ddattdef	C,D	[Construct] [Define Attribute]	dad	...

Command: **ddattdef**
Displays dialogue box:

COMMAND OPTIONS

Mode   Set the attribute text modes:
  Invisible   Make the attribute text invisible.
  Constant   Use constant values for the attributes.
  Verify   Verify the text after input.
  Preset   Preset the variable attribute text.

Attribute   Set the attribute text:
  Tag   Identifies the attribute.
  Prompt   Prompts the user for input.
  Value   Default value for the attribute.

Insertion point   Specify the attributes insertion point:
  Pick point   Pick insertion point with cursor.
  X   X-coordinate insertion point.
  Y   Y-coordinate insertion point.
  Z   Z-coordinate insertion point.

Text options	Specify the attribute text options:
Justification	Set the justification.
Text style	Select a style.
Height	Specify the height.
Rotation	Set the rotation angle.
Align	Automatically places the text below the previous attribute.

RELATED AUTOCAD COMMAND
- **AttDef**   Defines attribute definitions from the command line.

RELATED SYSTEM VARIABLES
- **AFlags**   Attribute mode:
  - 0   No mode specified.
  - 1   Invisible.
  - 2   Constant.
  - 4   Verify.
  - 8   Preset.
- **AttMode** Attribute display modes:
  - 0   Off.
  - 1   Normal.
  - 2   On.
- **AttReq**   Toggles prompt for attributes:
  - 0   Assume default values.
  - 1   Enables dialogue box or prompts for attributes.

DdAttE

Rel. 1

Edits attribute data via a dialogue box (*short for Dynamic Dialogue ATTribute Editor*).

Command	Alt+	Menu bar	Alias	GC Alias
ddatte	M,A	[Modify] [Edit Attribute]	de	...

```
Command: ddatte
Select block: [pick]
```
Displays dialogue box:

```
              Edit Attributes
Block name: COMPUTER

Type of Item:       [Computer          ]
Manufacturer:       [Touch             ]
Model:              [Pentium 60        ]
Cost:               [$3,105            ]
Purchase Date:      [17 January, 1995  ]
Barcode:            [010010111010100101]
Location:           [Room 101          ]
Employee:           [Ralph Grabowski   ]
Phone Extension:    [9597              ]
                    [                  ]

     [ OK ] [ Cancel ]  [Previous] [Next] [Help...]
```

COMMAND OPTIONS
None.

RELATED AUTOCAD COMMANDS
- **AttEdit**   Global attribute editor.
- **DdEdit**   Edits attribute definitions.

RELATED SYSTEM VARIABLE
- **AttDia**   Toggles use of **DdAttE** during **Insert** command.

TIP
- The **DdEdit** command edits attribute values (such as "Computer") but not attribute prompts, such as "Type of Item."

DdAttExt

Rel. 1

Extracts attribute information to a file (*short for Dynamic Dialogue ATTribute EXTraction*).

Command	Alt+	Menu bar	Alias	GC Alias
ddattext	F,I,A	[File] [Import/Export] [Attributes Out]	dax	...

Command: **ddattext**
Displays dialogue box:

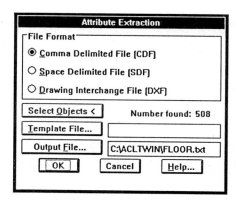

COMMAND OPTIONS
Comma delimited
    Create a CDF text file, where commas separate fields.
Space delimited Create a SDF text file, where spaces separate fields.
Drawing interchange
    Create an ASCII DXF file.
Select Entities Return to the graphics screen to select attributes for export.
Template file Specify the name of the TXT template file for CDF and SDF files.
Output File Specify the name of the attribute output file:
- filename.**TXT** for CDF and SDF formats.
- filename.**DXX** for DXF format.

RELATED AUTOCAD COMMAND
- **AttExt** Attribute extraction via command line interface.

DdChProp

Rel. 1

Modifies the color, layer, linetype, and thickness of most entities via a dialogue box (*short for Dynamic Dialogue CHange PROPerties*).

Command	Alt+	Menu bar	Alias	GC Alias
ddchprop	M,P	[Modify]	dc	...
		[Change Properties]		

Command: **ddchprop**
Select objects: **[pick]**
Displays dialogue box:

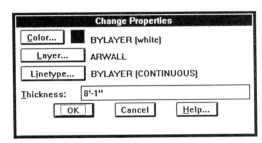

COMMAND OPTIONS

Color   Change the color of the selected entities by dialogue box:

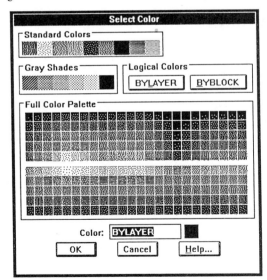

Layer name   Move the selected entities to a different layer by dialogue box:

[Select Layer dialog box showing Current Layer: 0, with layer list including 0, ARAPPLIANCE, ARCABINET, ARDOOR, ARDOORSPEC, ARGLAZING, ARHEADER, ARWALL (selected), ARWHEADER, ARWINDOWSPEC, DEFPOINTS, DIM, with State, Color, and Linetype columns. Set Layer Name: ARWALL, OK and Cancel buttons.]

LInetype   Change the linetype of the selected entities by dialogue box:

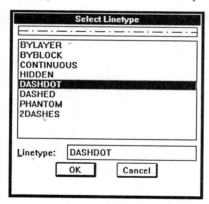

Thickness   Change the thickness of the selected entities.

RELATED AUTOCAD COMMANDS
- ChProp   Change properties via command line.
- Change   Allows changes to lines, circles, blocks, text, and attributes.

RELATED SYSTEM VARIABLES
- CeColor   The current entity color setting.
- CeLtype   The current entity linetype setting.
- CLayer   The name of the current layer.
- Thickness The current thickness setting.

 DdEdit

Rel. 1

Edits a single line of text or the tag, prompt, and default value of a single attribute using a dialogue box (*short for Dynamic Dialogue EDITor*).

Command	Alt+	Menu bar	Aliases	GC Aliases
ddedit	M,D	[Modify]	ed	ae
		[Edit Text]	te	te

Command: **ddedit**
<Select a TEXT or ATTDEF entity>/Undo: [pick text]
Displays dialogue box:

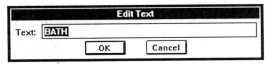

<Select a TEXT or ATTDEF entity>/Undo: [pick attribute definition]
Displays dialogue box:

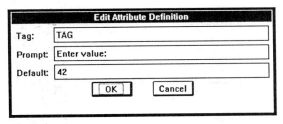

COMMAND OPTIONS
U   Undo editing operation.
[Ctrl]+C   Cancels to end the **DdEdit** command.

RELATED AUTOCAD COMMANDS
- **DdAttE**   Edits the values of text attributes connected with a block.
- **Change**   Edits some text attributes.

TIPS
- The **DdEdit** command automatically repeats; press **[Ctrl]+C** to cancel the command.
- Use the **DdAttE** command to edit the values of all attribute in a single block.

'DdEModes

Rel. 1

Sets the working parameters (*short for Dynamic Dialogue Entity MODES*).

Command	Alt+	Menu bar	Alias	GC Alias
'ddemodes	D,E	[Settings] [Entity Modes]	em	...

Command: **ddemodes**
Displays dialogue box:

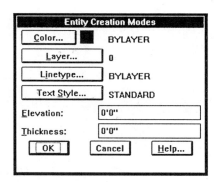

COMMAND OPTIONS

Elevation   Set the new working elevation.
Thickness   Set the new working thickness.
Layer   Set the new working layer.
Displays dialogue box:

```
                    Select Layer
Current Layer: 0
Layer Name              State       Color   Linetype
0                       On  . . . . white   CONTINUOUS
ARAPPLIANCE             On  . . . . 204     CONTINUOUS
ARCABINET               On  . . . . 42      CONTINUOUS
ARDOOR                  On  . . . . 42      CONTINUOUS
ARDOORSPEC              On  . . . . 142     CONTINUOUS
ARGLAZING               On  . . . . 42      CONTINUOUS
ARHEADER                On F . . . white    CONTINUOUS
ARWALL                  On  . . . . white   CONTINUOUS
ARWHEADER               On  . . . . 42      CONTINUOUS
ARWINDOWSPEC            On  . . . . 142     CONTINUOUS
DEFPOINTS               On  . . . . white   CONTINUOUS
DIM                     On  . . . . white   CONTINUOUS

Set Layer Name:  ARWALL
              OK        Cancel
```

DdEModes *cont'd*

Color   Set the new working color.
             Displays dialogue box:

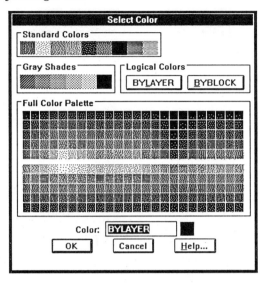

Linetype   Set the new working linetype.
             Displays dialogue box:

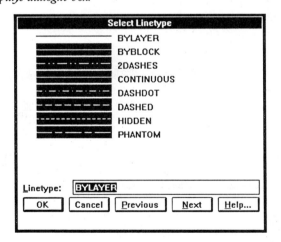

Text Style	Set the new working textstyle.
Displays dialogue box:

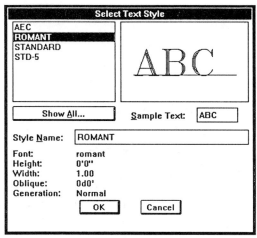

Show All	Displays all 255 characters in the selected font.
Displays dialogue box:

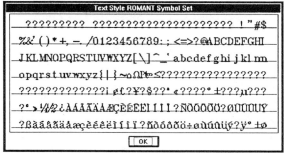

RELATED AUTOCAD COMMANDS
- Color   Set a new working color.
- Elevation   Set a new working elevation.
- Layer   Set a new working layer.
- Linetype   Set a new working linetype.
- Style   Set a new text style.

RELATED SYSTEM VARIABLES
- CeColor   The current entity color.
- CeLtype   The current entity linetype.
- CLayer   The current layer name.
- Elevation   The current elevation setting.
- TextStyle   The current text style setting.
- Thickness   The current thickness setting.

'DdGrips

Rel. 1

Turns entity grips on and off; defines the size and color of grips.

Command	Alt+	Menu bar	Alias	GC Alias
'ddgrips	S,G	[Settings] [Grips Style]	gr	...

Command: **ddgrips**
Displays dialogue box:

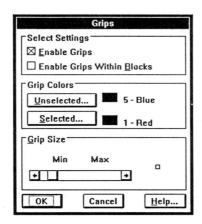

COMMAND OPTIONS

Enable grips Toggles the display of entity grips.
Enable grips within Blocks
  Displays grips on entities within blocks.
Grip size Change the size of the grip box.
Unselected Define the color of unselected grips; displays dialogue box.
Selected Define the color of selected grips.

52 ■ The Illustrated AutoCAD LT for Windows Quick Reference

Displays dialogue box:

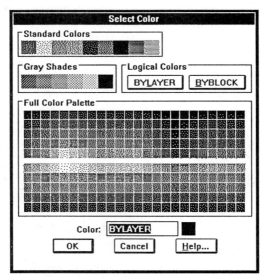

RELATED AUTOCAD COMMAND
- Select   Creates a selection set of entities.

RELATED SYSTEM VARIABLES
- Grips   Toggles use of grips:
  - 0   Disable grips.
  - 1   Enable grips (default).
- GripBlock   Toggles display of grips inside blocks:
  - 0   Display grip only in block insertion point (default).
  - 1   Display grips on entities inside block.
- GripColor   Color of unselected grips (default=5, blue).
- GripHot   Color of selected grips (default=1, red).
- GripSize   Size of grip, in pixels:
  - 1   Minimum size.
  - 3   Default size.
  - 255   Maximum size.

 # DDim

Rel. 1

Sets dimension styles and variables via a series of dialogue boxes (*short for Dialogue DIMension*).

Command	Alt+	Menu bar	Alias	GC Alias
ddim	D,M	[Settings] [Dimension Style]	dm	dd

Command: **ddim**
Displays dialogue box:

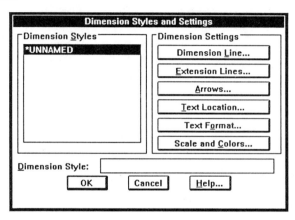

COMMAND OPTIONS

Dimension Styles Select a named dimension style.
Dimension Style Create a dimension style by naming it.
Dimension Line Specify the dimension line variables: toggle interior lines, and specify the text gap and baseline increment.

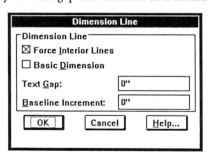

Extension Lines Specify the format of extension lines: location and visibility of extension line; and parameters of center mark.

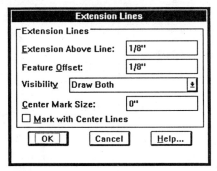

Arrows Specify the look of arrows: type, size, and visibility.

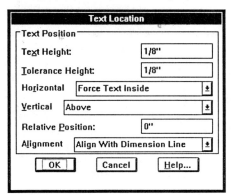

Text Location Specify the display of the dimensioning text.

DDim *cont'd*

Text Format   Specify the formatting of dimensioning text.

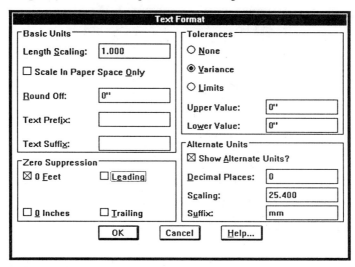

Scale and Colors Specify the overall dimension scale and the color of dimension elements.

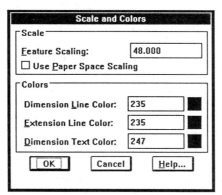

RELATED AUTOCAD COMMANDS
- *All dimensioning commands at the 'Dim:' prompt.*

RELATED SYSTEM VARIABLES
- **DimStyle**   Contains the name of the current dimension style.
- *All dimensioning variables.*

TIPS
- You cannot rename the default dimstyle named '*UNNAMED'.
- The current dimstyle name is stored in system variable **DimStyle**.
- You access dimstyles in externally referenced drawings via the **XBind** command.

DdInsert

Rel. 1

Insert blocks (symbols) via dialogue box (*short for Dialogue Dynamic INSERT*).

Command	Alt+	Menu bar	Alias	GC Alias
ddinsert	D,K	[Draw] [Insert Block]	i	...

Command: **ddinsert**
Displays dialogue box:

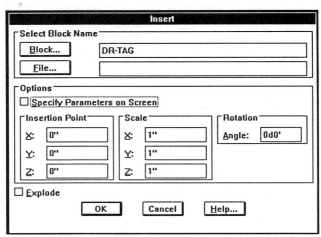

COMMAND OPTIONS

Block   Select the block name from a second dialogue box.
               Displays dialogue box:

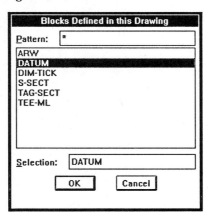

The Illustrated AutoCAD LT for Windows Quick Reference ■ 57

DdInsert *cont'd*

File   Select the drawing name from a second dialogue box.

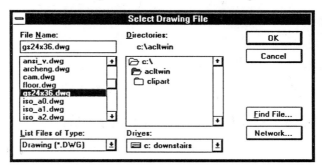

Specify parameters on screen
               Use the cursor to position the block.
Insertion point Specify the block's insertion point coordinates.
Scale   Specify the block's scale.
Rotation   Specify the block's angle of rotation.
Explode   Insert the block as individual entities.

RELATED AUTOCAD COMMANDS
- **Base**   Redefine the insertion point relative to the origin.
- **Block**   Create a block from a group of entities.
- **BMake**   Create a block using a dialogue box.
- **Explode** Explodes a block after insertion.
- **Insert**   Insert blocks via the command line.

RELATED SYSTEM VARIABLES
- **InsName** Default name for most-recently inserted block.
- **InsBase** Last setting of the **Base** command (default = 0,0,0).

'DdLModes

Rel. 1

Controls the layer settings in the drawing via a dialogue box (*short for Dynamic Dialogue Layer MODES*).

Command	Alt+	Menu bar	Alias	GC Alias
'ddlmodes	S,L	[Settings] [Layer Control]	ld	...

Command: **ddlmodes**
Displays dialogue box:

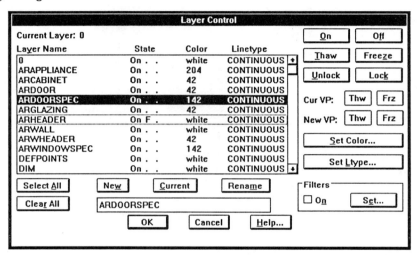

COMMAND OPTIONS

Clear all	Clear all selected layers.
Current	Set the working (current) layer name.
Freeze	Freeze the selected layers.
Lock	Lock the selected layers.
New	Create a new layer name.
Off	Turn off the selected layers.
On	Turn on the selected layers.
Rename	Rename a layer.
Select All	Select all layers.
Set Color	Set a new color for the selected layers.
Set Ltype	Set a new linetype for the selected layers.
Thaw	Thaw the selected frozen layers.
Unlock	Unlock the selected locked layers.

DdLModes *cont'd*

Set filters   Create a filter set via dialogue box:

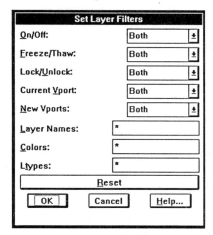

RELATED AUTOCAD COMMANDS
- **Layer**   Controls the current layer setting.
- **Rename**   Renames a layer.
- **VpLayer** Controls the layer settings in paper space.

RELATED SYSTEM VARIABLE
- **CLayer**   The current layer setting.

TIPS
- When a layer has been turned:
    - **Off**   AutoCAD no longer displays nor plots entities on that layer.
    - **Frozen** AutoCAD no longer takes its entities into account during a regeneration; in addition, AutoCAD no longer displays nor plots entities on that layer.
    - **Locked** Entities are displayed and you can draw on the layer but you cannot edit the layer's entities.
- Freezing a layer is more efficient than turning the layer off.

- Locking layers is useful for redlining; you can make additions and notes but not change the drawing.

DdModify

Rel. 2

Views and edits the properties of almost all entities via a dialogue box.

Command	Alt+	Menu bar	Alias	GC Alias
ddmodify	M,F	[Modify] [Modify Entity]	...	tg

```
Command: ddmodify
Select object to list: [pick object]
```
A different dialogue box appears for type of object.

COMMAND OPTIONS

Color	Change the object's color via a dialogue box.
Layer name	Move the entity to a different layer.
Linetype	Change the object's linetype via dialogue box.
Thickness	Change the object's thickness.
[pick arc]	Displays the **Modify Arc** dialogue box.

Displays dialogue box:

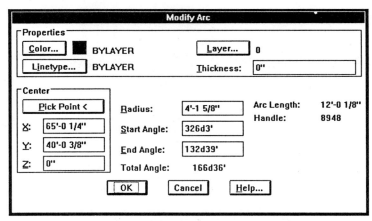

[pick hatch]   Displays the **Modify Associative Hatch** dialogue box.
Displays dialogue box:

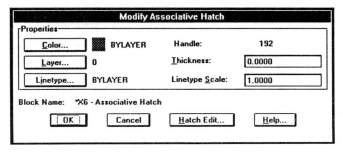

DdModify *cont'd*

[pick attribute] Displays the **Modify Attribute Definition** dialogue box.
Displays dialogue box:

[pick block] Displays the **Modify Block Insertion** dialogue box.
Displays dialogue box:

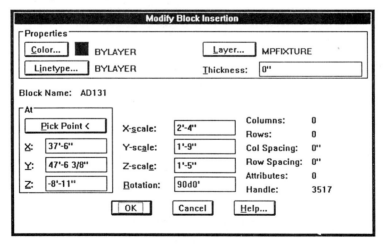

62 ■ The Illustrated AutoCAD LT for Windows Quick Reference

[pick circle]   Displays the **Modify Circle** dialogue box.
Displays dialogue box:

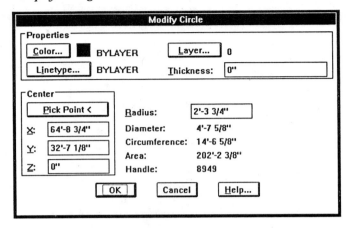

[pick line]   Displays the **Modify Line** dialogue box.
Displays dialogue box:

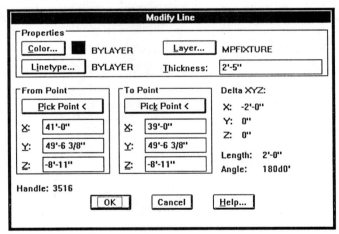

DdModify *cont'd*

[pick point]   Displays the **Modify Point** dialogue box.
Displays dialogue box:

[pick polyline]   Displays the **Modify Polyline** dialogue box.
Displays dialogue box:

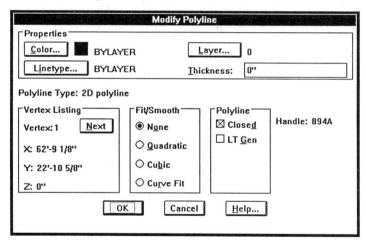

[pick solid]   Displays the **Modify Solid** dialogue box.
Displays dialogue box:

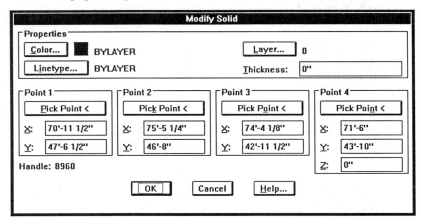

[pick text]   Displays the **Modify Text** dialogue box.
Displays dialogue box:

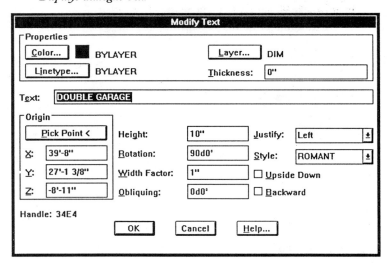

DdModify *cont'd*

[pick xref] Displays the **Modify External Reference** dialogue box.
Displays dialogue box:

```
┌─────────────── Modify External Reference ───────────────┐
│ ┌─Properties─────────────────────────────────────────┐  │
│ │  [Color...] ■ BYLAYER      [Layer...] 0            │  │
│ │  [Linetype...] BYLAYER     Thickness: [0.0000]     │  │
│ └────────────────────────────────────────────────────┘  │
│  Xref Name: XREF1        Path: XREF1.DWG                │
│ ┌─At─────────────────────────────────────────────────┐  │
│ │  [Pick Point <]   X-scale: [1.0000]  Columns:  0   │  │
│ │  X: [0.0000]      Y-scale: [1.0000]  Rows:     0   │  │
│ │  Y: [0.0000]      Z-scale: [1.0000]  Col Spacing: 0.0000 │
│ │  Z: [0.0000]      Rotation: [0]      Row Spacing: 0.0000 │
│ │                                      Handle:  None │  │
│ └────────────────────────────────────────────────────┘  │
│            [  OK  ]   [ Cancel ]   [ Help... ]          │
└─────────────────────────────────────────────────────────┘
```

RELATED AUTOCAD COMMANDS
- **Change**   Changes most properties of entities.
- **ChProp**   Changes some properties of entities.
- **DdChProp** Edits properties via dialogue box.
- **DdEdit**   Edits text via dialogue box.
- **PEdit**   Edits polylines and meshes.

RELATED SYSTEM VARIABLES
- *Many system variables.*

TIPS
- **DdModify** does not work with the following objects in an AutoCAD LT drawing:
    - Associative hatch.
    - Associative dimension.
    - Shape.
    - Trace.
    - Viewport.
    - 3D objects, such as 3D face.
    - Solids modelling, such as AME objects.

- For hatches, **DdModify** displays the **Modify Block Insertion** dialogue box.

- For polygons and ellipses, **DdModify** displays the **Modify Polyline** dialogue box.

 'DdOsnap

Rel. 1

Sets entity snap modes and aperture size via dialogue box (*short for Dynamic Dialogue Entity SNAP*).

Command	Alt+	Menu bar	Alias	GC Alias
'ddosnap	A,O	[Assist] [Object Snap]	os	...

Command: **ddosnap**
Displays dialogue box:

COMMAND OPTIONS
Select settings Select the entities snaps.
Aperture size Change the size of the aperture.

RELATED AUTOCAD COMMANDS
- **OSnap** Sets entity snap modes via the command line.
- **<middle button>**
  Displays list of entity snap modes.

RELATED SYSTEM VARIABLES
- **Aperture** Size of the entity snap aperture.
- **OsMode** The current entity snap modes.

TIP
- See the **OSnap** command for a visual list of object snaps.

'DdPtype

Rel. 1

Sets the type and size of point entities via a dialogue box (*short for Dynamic Dialogue Point TYPE*).

Command	Alt+	Menu bar	Alias	GC Alias
'ddptype	S,O	[Settings] [Point Style]	...	...

Command: **ddptype**
Displays dialogue box:

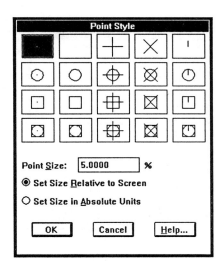

COMMAND OPTIONS
Point size   Set size in percent or pixels.
Set size Relative to screen
    Set size in percent.
Set size in Absolute units
    Set size in pixels.

RELATED AUTOCAD COMMAND
- Point   Draws points.

RELATED SYSTEM VARIABLES
- PdMode   Determines the look of a point.
- PdSize   Contains the size of the point.

DdRename

Rel. 1

Changes the names of blocks, dimension styles, layers, linetypes, text styles, named UCSs, view names, and viewports via a dialogue box.

Command	Alt+	Menu bar	Alias	GC Alias
ddrename	M,N	[Modify] [Rename]	dr	...

Command: **ddrename**
Displays dialogue box:

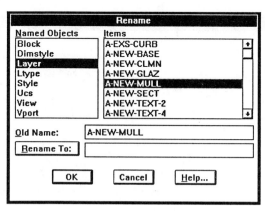

COMMAND OPTIONS
Named Objects Select type of named object.
Items Select named object.
Old name Specify the name (or group of names) to change.
Rename to Indicate the new name.

RELATED AUTOCAD COMMANDS
- **DdModify** Changes names (and all other attributes) of entities.
- **Rename** Changes the names of entities via the command line.

TIPS
- You cannot rename layer '0', dimstyle 'Unnamed', anonymous blocks, groups, and linetype 'Continuous'.

- To rename a group of similar names, use * (the wildcard for "all") and ? (the wildcard for a single character).

- Names can be up to 31 characters in length, including the $, -, and _ characters.

'DdRModes

Rel. 1

Controls the current settings of snap, snap angle, grid, axes, ortho, blips, and isometric modes (*short for Dynamic Dialogue dRawing MODES*).

Command	Alt+	Menu bar	Alias	GC Alias
'ddrmodes	S,D	[Settings] [Drawing Aids]	da	...

Command: **ddrmodes**
Displays dialogue box:

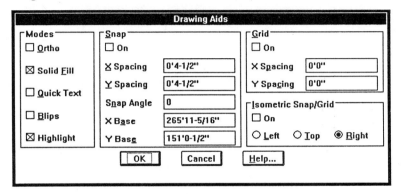

COMMAND OPTIONS

Ortho	Turn orthographic mode on and off.
Solid Fill	Turn solid fill on and off.
Quick Text	Turn quick text on and off.
Blips	Turn blipmarks on and off.
Highlight	Turn entity highlighting on and off.
Snap	Turn snap mode on and off:

  Snap X spacing
        Set x-spacing for snap.
  Snap Y spacing
        Set y-spacing for snap.
  Snap angle Set angle for snap and grid.
  X Base Set snap, grid hatch x-basepoint.
  Y base Set snap, grid hatch y-basepoint.

Grid Turn grid marks on and off:
  X spacing Set grid x-spacing.
  Y spacing Set grid y-spacing.

Isometric	Turn isometric mode on and off:
Left	Switch to left isometric plane.
Top	Switch to top isometric plane.
Right	Switch to right isometric plane.

RELATED AUTOCAD COMMANDS

- **Blipmode** — Toggles visibility of blip markers.
- **Fill** — Toggles fill mode.
- **Grid** — Sets the grid spacing and toggles visibility.
- **Highlight** — Toggles highlight mode.
- **Isoplane** — Selects the working isometric plane.
- **Ortho** — Toggles orthographic mode.
- **QText** — Toggles quick text mode.
- **Snap** — Sets the snap spacing and isometric mode.

RELATED SYSTEM VARIABLES

- **BlipMode** — Current blip marker visibility:
  0 Off.
  1 On
- **FillMode** — Current fill mode:
  0 Off.
  1 On.
- **GridMode** — Current grid visibility:
  0 Off.
  1 On.
- **GridUnit** — Current grid spacing.
- **Highlight** — Current highlight mode:
  0 Off.
  1 On.
- **OrthoMode** — Current orthographic mode setting:
  0 Off.
  1 On.
- **QTextMode** — Current quick text mode setting:
  0 Off.
  1 On.
- **PickStyle** — Controls group selection:
  0 Groups and associative hatches not selected.
  1 Groups selected.
  2 Associative hatches selected.
  3 Both selected.
- **SnapAng** — Current snap and grid rotation angle.

DdRModes *cont'd*

- **SnapBase** Base point of snap and grid rotation angle.
- **SnapIsoPair** Current isoplane:
    - 0 Left.
    - 1 Top.
    - 2 Right.
- **SnapMode** Current snap mode setting:
    - 0 Off.
    - 1 On.
- **SnapStyl** Snap style setting:
    - 0 Standard.
    - 1 Isometric.
- **SnapUnit** Current snap spacing.

TIPS

- **DdRModes** is an alternative to the **SetVar** command for checking the status of the above 14 system variables.

- Use the function key **F7** (or control key **[Ctrl]+G**) to turn the grid on and off during a command.

- Use the function key **F8** (or control key **[Ctrl]+O**) to change ortho mode during a command.

- Use the function key **F9** (or control key **[Ctrl]+B**) to change snap mode during a command.

'DdSelect

Rel. 1

Defines the type of entity selection mode and pickbox size via a dialogue box.

Command	Alt+	Menu bar	Alias	GC Alias
'ddselect	S,C	[Settings] [Selection Style]	sl	...

Command: **ddselect**
Displays dialogue box:

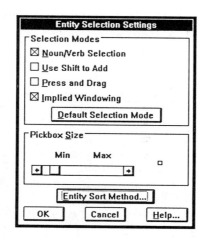

COMMAND OPTIONS

Selection modes Specify the style of entity selection mode:
   Noun/verb   Select entities first, then enter the command.
   Use shift to add
         The **[Shift]** key adds entities to selection set.
   Press and drag Create windowed selection set by pressing mouse key and dragging window, rather than specifying two points.
   Implied windowing
         Automatically create a windowed selection box.
Default selection mode
         Reset modes to turn Noun/verb and Implied windowing on.
Pickbox size Interactively change the size of the pickbox.

DdSelect *cont'd*

Entity sort method
: Displays a second dialogue box to specify the commands that sort entities by drawing database order.
Displays dialogue box:

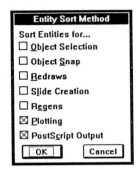

Object Selection
: Entities are added to the selection in database order.
Entity Snap
: Entity snap modes find entities in database order.
Redraws
: Redraws entities in database order.
Slide Creation
: The **MSlide** command draws entities in database order.
Regens
: Regenerates entities in database order.
Plotting
: The **Plot** command processes entities in database order.
PostScript Output
: The **PsOut** command processes entities in database order.

RELATED AUTOCAD COMMAND
Select   Creates a selection set before executing an editing commands.

RELATED SYSTEM VARIABLES
- **PickAdd**   Determines effect of **[Shift]** key on creating selection set:
  - 0   [Shift] key adds to selection set.
  - 1   [Shift] key removes from selection set (default).
- **PickAuto**   Determines automatic windowing:
  - 0   Disabled.
  - 1   Enabled (default).
- **PickBox**   Specifies the size of the pickbox.
- **PickDrag**   Method of creating selection window:
  - 0   Click at both corners (default).
  - 1   Click one corner, drag to second corner.
- **PickFirst** Method of entity selection:
  - 0   Enter command first.
  - 1   Select entities first (default).

- **SortEnts**   Entities are displayed in database order during:
  - 0   Off.
  - 1   Entity selection.
  - 2   Entity snap.
  - 4   Redraw.
  - 8   Slide generation.
  - 16 Regeneration.
  - 32 Plots.
  - 64 PostScript output.

TIPS

- These commands work with noun-verb selection: **Array, Block, Change, ChProp, Copy, DdChProp, DView, Erase, Explode, Hatch, List, Mirror, Move, Rotate, Scale, Stretch,** and **WBlock.**

- A larger pickbox makes it easier to select entities but also makes it easier to inadvertently select entities.

- Use **Entity Sort Method** if the drawing requires that entity be processed in the order they appear in the drawing, such as for NC applications.

- **Plotting** and **PostScript Output** are turned on, by default; setting more sort methods increases processing time.

# DdUcs

Rel. 1

Creates and controls UCS planes via a dialogue box (*short for Dynamic Dialogue User Coordinate System*).

Command	Alt+	Menu bar	Alias	GC Alias
dducs	A,N	[Assist] [Named UCS]	uc	...

Command: **dducs**
Displays dialogue box:

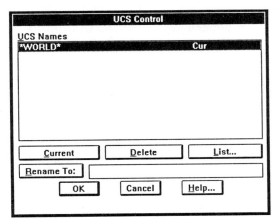

COMMAND OPTIONS

Current   Make the selected name the current UCS.
Delete   Delete a named UCS.
Rename to   Rename a UCS.
List   Lists information about the selected UCS in a dialogue box:

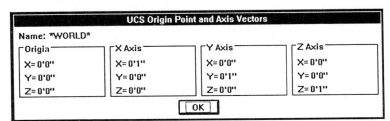

RELATED AUTOCAD COMMANDS

- DdUcsP   Selects a predefined UCS from a dialogue box.
- Ucs   Creates and saves user-defined coordinate systems.

RELATED SYSTEM VARIABLES
- **UcsFollow**   New UCS is displayed in plan view.
- **UcsName**   Current name of UCS.
- **UcsOrg**   WCS origin of the current UCS.
- **UcsXdir**   X-direction of the current UCS.
- **UcsYdir**   Y-direction of the current UCS.
- **ViewMode**   Current clipped viewing mode.
- **WorldUcs**   UCS = WCS toggle.
- **WorldView** UCS or WCS for **Dview** and **Vpoint** commands.

TIP
- Create a UCS to make it easier to draw in three dimensions.

DdUcsP

Rel. 1

Selects one of nine predefined user coordinate systems (*short for Dynamic Dialogue UCS Preset*).

Command	Alt+	Menu bar	Alias	GC Alias
dducsp	A,P	[Assist] [Preset UCS]	up	...

Command: **dducsp**
Displays dialogue box:

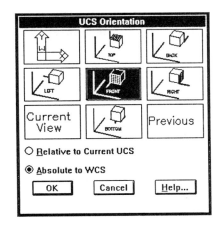

COMMAND OPTIONS

Relative to current UCS
  Switch to preset UCS relative to current UCS.
Absolute to WCS
  Switch to preset UCS relative to WCS (world coordinate system).

RELATED AUTOCAD COMMANDS

- **DdUcs**   Creates and selects named UCS.
- **Ucs**   Sets the current UCS via the command line.

RELATED SYSTEM VARIABLES

- **UcsFollow** New UCS is displayed in plan view.
- **UcsName** Current name of UCS.
- **UcsOrg** WCS origin of the current UCS.
- **UcsXdir** X-direction of the current UCS.
- **UcsYdir** Y-direction of the current UCS.
- **ViewMode** Current clipped viewing mode.
- **WorldUcs** UCS = WCS toggle.
- **WorldView** UCS or WCS for **DView** and **VPoint** commands.

'DdUnits

Rel. 1

Select the display of units and angles via a dialogue box.

Command	Alt+	Menu bar	Alias	GC Alias
'ddunits	S,U	[Settings]	du	...
		[Unit Style]		

Command: **ddunits**
Displays dialogue box:

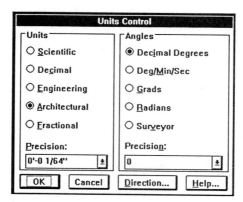

COMMAND OPTION

Direction   Displays a second dialogue box to specify the direction of 0 degrees:

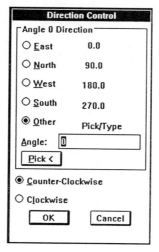

RELATED AUTOCAD COMMAND

- Units   Sets units and angles via the command line.

DdUnits *cont'd*

RELATED SYSTEM VARIABLES

- **AngBase**   Direction of zero degrees relative to the current UCS.
- **AngDir**   Direction of angle measurement:
    - 0 Clockwise.
    - 1 Counterclockwise.
- **AUnits**   Style of angle units:
    - 0 Decimal degrees.
    - 1 Degree-minutes-seconds.
    - 2 Grads.
    - 3 Radians.
    - 4 Surveyor's units.
- **AuPrec**   Decimal places of angle units
- **LUnits**   Style of linear units:
    - 0 Scientific.
    - 1 Decimal.
    - 2 Engineering.
    - 3 Architectural.
    - 4 Fractional.
- **LuPrec**   Decimal places of linear units.
- **ModeMacro** Customizes the status line via the Diesel language.
- **UnitMode** Displays input units:
    - 0 As set by **DdUnits** or **Units** command (the default).
    - 1 As input by the user.

TIPS

- Distance formats:
    - **Decimal:**   0.0000 (default)
    - **Architectural:** 0'-0/64" (feet and fractional inches)
    - **Engineering:**   0'-0.0000" (feet and decimal inches)
    - **Fractional:**   0 0/64 (unitless fractional)
    - **Scientific:**   0.0000E+01

- Angular formats:
    - **Decimal:**   0.0000 (default)
    - **Deg-Min-Sec:**   0d0'0.0000" (degrees, minutes, decimal seconds)
    - **Radian:**   0.0000r
    - **Grad:**   0.0000g
    - **SurveyorUnits:** N0d'0.0000"E

'Delay

Rel. 1

Delays the next script command, in milliseconds.

Command	Alt+	Menu bar	Alias	GC Alias
'delay	...	...	...	...

```
Command: delay
Delay time in milliseconds:
```

COMMAND OPTIONS
None.

RELATED AUTOCAD COMMAND
- Script   Initiates a script.

TIPS
- Use the **Delay** command to slow down the execution of a script file.
- The maximum delay is 32767, just over 32 seconds.

Dim

Rel. 1

Changes the prompt from 'Command:' to 'Dim:'; allows access to AutoCAD's dimensioning commands (*short for DIMension*).

Command	Alt+	Menu bar	Alias	GC Alias
dim	D,I	[Draw] [Linear Dimensions]	d	lx
	D,N	[Draw] [Ordinate Dimensions]	...	...
	D,M	[Draw] [Radial Dimensions]	...	ix rx
	D,U	[Draw] [Angular Dimension]	...	...
	D,R	[Draw] [Leader]	...	le

Command: **dim**
Dim:

COMMAND OPTIONS

(Aliases for the dimension commands are shown in uppercase.)

■ **Commands for drawing dimensions:**
ALigned   Draws linear dimension aligned with object.
ANgular   Draws angular dimension that measures an angle.
Baseline   Continues a dimension from a basepoint.
CEnter   Draws a centermark on circle and arc centers.
COntinue   Continues from the previous dimension's second extension line.
Diameter   Draws diameter dimension on circles, arcs, and polyarcs.
HORizontal   Draws a horizontal dimension.
LEAder   Draws a leader.
ORdinate   Draws x- and y-ordinate dimensions.
RAdius   Draws radial dimension on circles, arcs, and polyline arcs.
VErtical   Draws vertical linear dimensions.

■ **Commands for editing dimensions:**
HOMetext   Returns associative dimension text to its original position.
Newtext   Edits text in associative dimensions.
OBlique   Changes angle of extension lines in associative dimensions.
REStore   Restores a dimension to the current dimstyle.
ROtated   Draws a linear dimension at any angle.
TEdit   Changes location and orientation of text in associative dimensions.
TRotate   Changes the rotation of text in associative dimensions.
Undo   Undoes the last dimension action.
UPdate   Updates selected associative dimensions to the current dimvar settings.

- **Commands for utility functions:**

Exit	Returns from 'Dim:' prompt to 'Command:' prompt.
OVerride	Overrides the current set of dimension variables.
REDraw	Redraws the current viewport (same as 'Redraw).
SAve	Saves the current setting of dimension styles as a dimstyle.
STAtus	Lists the current settings of dimension variables.
STYle	Sets a text style for the dimension text.
VAriables	Lists values of variables associated with a dim style (*not* dimvars).

RELATED AUTOCAD COMMANDS

- **DDim**   Dialogue box for setting dimension variables.
- **Dim1**   Executes a single dimensioning command, then returns to the 'Command:' prompt.
- **Style**   Determines the text style of the dimensioning text.
- **Units**   Determines the angular and linear styles of dimensioning text.

RELATED SYSTEM VARIABLES
All dimension related system variables.

RELATED AUTOCAD BLOCK
- **Dot**   AutoCAD uses a dot in place of the arrow head.

TIPS
- Only transparent commands and dimension commands work at the 'Dim:' prompt. To use other commands, you must exit the 'Dim:' prompt back to the 'Command:' prompt with the **Exit** comand.

- Most dimensions consists of four basic components: dimension line, extension lines, arrowheads, and text, as shown below.

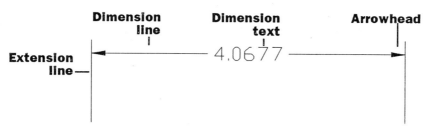

- All the components of an *associative dimension* are treated as a single object; components of a non-associative dimension are treated as individual objects.

Dim1

Rel. 1

Executes a single dimensioning command, returns to the 'Command:' prompt (*short for DIMension once*).

Command	Alt+	Menu bar	Alias	GC Alias
dim1	...	...	...	...

Command: **dim1**
Dim:

COMMAND OPTIONS
All dimension commands; see Dim command for list.

RELATED AUTOCAD COMMANDS
- **DDim** Dialogue box for setting dimension variables.
- **Dim** Switches to dimensioning mode and remains there.

RELATED SYSTEM VARIABLES
- **DimAso** Determines whether dimensions are drawn associatively.
- **DimTxt** Determines the height of text.
- **DimScale** Determines the dimension scale.

RELATED AUTOCAD BLOCK
- **Dot** Dim uses a dot in place of the arrow head.

TIP
- Use the **Dim1** command when you need to create or edit a single dimension command.

Dim: Aligned

Rel. 1

Draws linear dimensions aligned with an object.

Dim	Alt+	Menu bar	Alias	GC Alias
aligned	D,I,A	[Draw] [Linear Dimensions] [Aligned]	d al	lx

Dim: **aligned**
First extension line origin or RETURN to select: **[Enter]**
Select object to dimension: **[pick]**
Dimension line location (Text/Angle): **T**
Dimension text <>: **[Enter]**

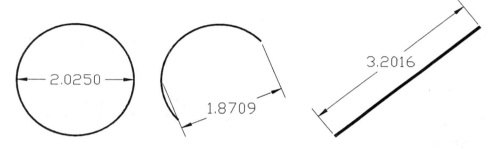

COMMAND OPTIONS
[Enter] Displays the submenu for selecting entities:
  [pick line] Dimensions line.
  [pick arc] Dimensions arc.
  [pick circle] Dimensions circle.
  [pick pline] Dimensions an individual segment.
Text Specify dimension text.
Angle Indicate dimension angle.

RELATED DIM COMMAND
n Rotated Draws angular dimension line with perpendicular extension line.

RELATED SYSTEM VARIABLE
n DimExo Dimension line offset distance.

Dim: Angular

Rel. 1

Draws a dimension that measures an angle.

Dim	Alt+	Menu bar	Alias	GC Alias
angular	D,U	[Draw] [Angular Dimension]	d an	...

Dim: **dimangular**
Select arc, circle, line, or RETURN: **[pick]**
Second angle endpoint: **[pick]**
Dimension line arc location (Text/Angle): **T**
Dimension text <>: **[Enter]**

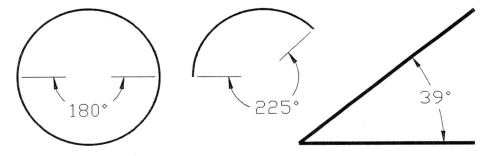

COMMAND OPTIONS
[pick arc]   Measures the angle of the arc.
[pick circle] Prompts you to pick two points on the circle.
[pick line]   Prompts you to pick two lines.
[Enter]   Prompts you to pick points to make an angle.
Text   Specify dimension text.
Angle   Indicate dimension angle.

Dim: Baseline

Rel. 1

Draws linear dimension from the previous starting point.

Dim	Alt+	Menu bar	Alias	GC Alias
baseline	D,I,B	[Draw] [Linear Dimensions] [Baseline]	d b	...

Dim: `baseline`
Second extension line origin or RETURN to select: [pick]
Dimension text:

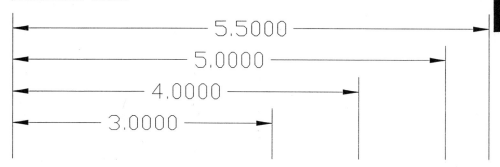

Original Dimension **Baseline Dimensions**

COMMAND OPTIONS
[Enter]   Prompts you to select the original dimension.

RELATED DIM COMMAND
- Continue   Continues linear dimensioning from last extension point.

RELATED SYSTEM VARIABLES
- DimDli   Specifies the distance between baseline dimension lines.
- DimSe1   Suppress first extension line.

Dim: Center

Rel. 1

Draws center lines on arcs and circles.

Dim	Alt+	Menu bar	Alias	GC Alias
center	D,M,C	[Draw]	d ce	. . .
		[Radial Dimensioning]		
		[Center Mark]		

Dim: **center**
Select arc or circle: **[pick]**

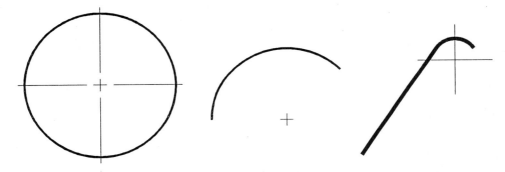

COMMAND OPTIONS
None.

RELATED DIM COMMANDS
- **Aligned** Dimensions arcs and circles.
- **Diameter** Dimensions arcs and circles by diameter value.
- **Radius** Dimensions arcs and circles by radius value.

RELATED SYSTEM VARIABLE
- **DimCen** Size of the center mark:
  0.09 Draws center mark only (default value).
  - 0.09 Draws center mark with extension lines.

Dim: Continue

Rel. 1

Continues a dimension from the second extension line of the previous dimension.

Dim	Alt+	Menu bar	Alias	GC Alias
continue	D,I,C	[Draw] [Linear Dimensions] [Continue]	d co	...

```
Dim: continue
Select continued dimension: [pick]
Second extension line origin or RETURN to select: [pick]
Dimension text:
```

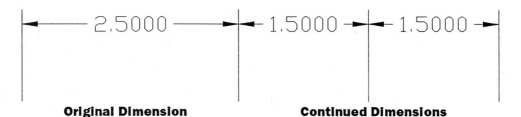

      Original Dimension   **Continued Dimensions**

COMMAND OPTIONS
[Enter]   Prompts you to select originating dimension.

RELATED DIM COMMAND
- Baseline   Continue dimensioning from first extension point.

RELATED SYSTEM VARIABLES
- DimDli   Distance between continuous dimension lines.
- DimSe1   Suppress first extension line.

Dim: Diameter

Rel. 1

Draws a diameter dimension on arcs, circles, and polyline arcs.

Dim	Alt+	Menu bar	Alias	GC Alias
diameter	D,M,D	[Draw]	d d	ix
		[Radial Dimensions]		
		[Diameter]		

```
Dim: diameter
Select arc or circle: [pick]
Dimension text: [Enter]
Enter leader length for text: [pick]
```

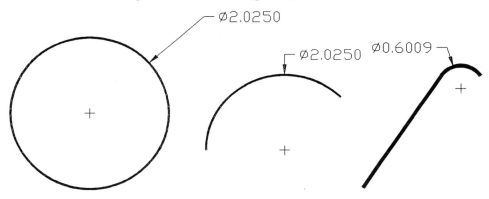

COMMAND OPTIONS
None.

RELATED DIM COMMANDS
- Center   Marks the center point of arcs and circles.
- Radius   Draws the radius dimension of arcs and circles.

RELATED SYSTEM VARIABLE
- DimCen   Controls the size of the center mark.

Dim: Exit

Rel. 1

Exits the 'Dim:' prompt back to the 'Command:' prompt.

Dim	Alt+	Menu bar	Alias	GC Alias
exit	...	...	d e	...

Dim: **exit**
Command:

COMMAND OPTIONS
None.

RELATED AUTOCAD COMMANDS
- **Dim**   Changes the 'Command:' prompt to the 'Dim:' prompt.
- **Dim1**   Switches to the dimensioning mode for just one dimension command.

RELATED SYSTEM VARIABLE
None.

TIP
- Using the **Undo** (or **U**) command after returning to the 'Command:' prompt undoes all dimensioning performed while at the 'Dim:' prompt.

Dim: Hometext

Rel. 1

Returns the text of an associative dimension to its original position.

Dim	Alt+	Menu bar	Alias	GC Alias
hometext	M,I,H	[Modify] [Edit Dimension] [Home Position]	d hom	...

```
Dim: hometext
Select objects: [pick]
Select objects: [Enter]
```

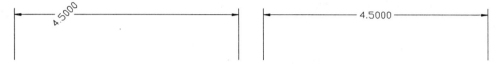

Rotated, moved dimension text   **Hometext applied**

COMMAND OPTIONS
Select objects   Selects the dimensions with text to be returned to the home position.

RELATED DIM COMMAND
- **TEdit**   Changes the location of text in associative dimensions.

RELATED SYSTEM VARIABLE
- **DimAso**   Controls whether dimensions are associative.

Dim: Horizontal

Rel. 1

Draws horizontal dimensions.

Command	Alt+	Menu bar	Alias	GC Alias
horizontal	D,I,H	[Draw]	d hor	lx
		[Linear Dimensions]		
		[Horizontal]		

```
Dim: horizontal
First extension line origin or RETURN to select: [Enter]
Select line, arc or circle: [pick]
Dimension line location(Text/Angle): T
Dimension text <>: [Enter]
```

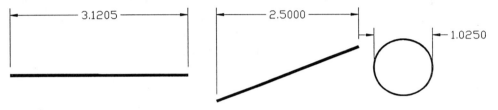

COMMAND OPTIONS

[Enter]   Displays the submenu for dimensioning objects:
    Select line, arc or circle
Text   Allows you to change the text of the dimension.
Angle   Specifies the angle of the text.

RELATED DIM COMMAND
- **Aligned**   Draws linear dimensions aligned with objects.
- **Vertical**   Draws vertical dimensions.

Dim: Leader

Rel. 1

Draws leader line with a single line of text.

Command	Alt+	Menu bar	Alias	GC Alias
leader	D,L	[Draw] [Leader]	d lea	le

```
Dim: leader
Leader start: [pick]
To point: [pick]
To point: [Enter]
Dimension text <>:
```

COMMAND OPTIONS
Leader start   Starting point of the leader line.
To point   Indicate vertex (corner) of leader line; press **[Enter]** to end line drawing.
Dimension text
   Specify the leader text.

RELATED DIM COMMANDS
None.

Dim: Newtext

Rel. 1

Edits text in associative dimensions.

Command	Alt+	Menu bar	Alias	GC Alias
newtext	M,I,C	[Modify] [Edit Dimension] [Change Text]	d n	...

```
Dim: newtext
Enter new dimension text:
Select objects: [pick]
Select objects: [Enter]
```

Example usage:
```
Dim: newtext
Enter new dimension text: Adjust to fit
Select objects: [pick]
Select objects: [Enter]
```

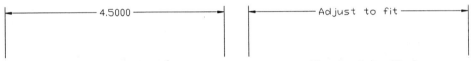

Original Dimension   **Newtext applied**

COMMAND OPTION
Enter new text   Type new text for dimension.

RELATED DIM COMMANDS
None.

Dim: Oblique

Rel. 1

Changes the angle of the extension lines of associative dimensions.

Command	Alt+	Menu bar	Alias	GC Alias
oblique	M,I,O	[Modify] [Edit Dimension] [Oblique Dimension]	d ob	...

```
Dim: oblique
Select objects: [pick]
Select objects: [Enter]
Enter obliquing angle (RETURN for none):
```

Obliqued extension lines

COMMAND OPTION
[Enter]   Return oblique extension lines back to upright.

RELATED DIM COMMAND
- TRotate   Rotates text in associative dimensions.

Dim: Ordinate

Rel. 1

Draws an x- or y-ordinate dimension.

Dim	Alt+	Menu bar	Alias	GC Alias
ordinate	D,N	[Draw] [Ordinate Dimensions]	d or	...

```
Command: dimordinate
Select Feature: [pick]
Leader endpoint (Xdatum/Ydatum/Text): [pick]
Leader endpoint: [pick]
```

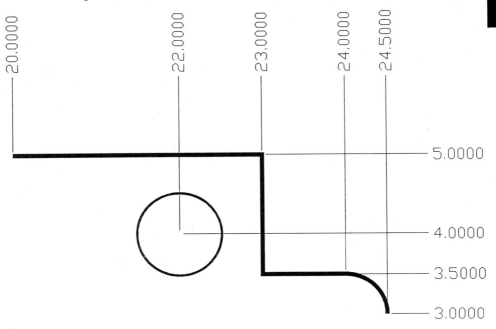

COMMAND OPTIONS
Text   Places text, rather than dimension.
Xdatum   Forces x-ordinate dimension.
Ydatum   Forces y-ordinate dimension.

RELATED DIM COMMANDS
- Horizontal   Draws regular horizontal dimensions.
- Leader   Draws leader dimensions.
- Vertical   Draws regular vertical dimensions.

Dim: Override

Rel. 1

Overrides the currently set dimension variables.

Dim	Alt+	Menu bar	Alias	GC Alias
override	...	...	d ov	...

```
Dim: override
Dimension variable to override:
Current value xxx New Value:
Select objects: [pick]
```

COMMAND OPTIONS
None.

RELATED DIM COMMAND
- **Style**   Creates and modifies dimension styles.

RELATED SYSTEM VARIABLES
- *All dimension variables.*

Dim: Radius

Rel. 1

Draws radial dimensions on circles, arcs, and polyline arcs.

Dim	Alt+	Menu bar	Alias	GC Alias
radius	D,M,R	[Draw] [Radial Dimensions] [Radius]	d ra	rx

```
Dim: radius
Select arc or circle: [pick]
Dimension line location (Text/Angle): T
Dimension text <>:
```

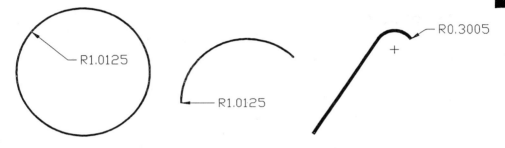

COMMAND OPTIONS
Text   Place text.
Angle   Change angle of dimension text.

RELATED DIM COMMANDS
- Center   Draws center mark on arcs and circles.
- Diameter Draws diameter dimensions on arcs and circles.

RELATED SYSTEM VARIABLE
- DimCen Determines the size of the center mark.

TIP
To include the diameter symbol, use the text code: %%d .

Dim: Redraw

Rel. 1

Redraws the drawing to clean up the window.

Dim	Alt+	Menu bar	Alias	GC Alias
redraw	...	...	d red	rd

Dim: `redraw`

COMMAND OPTIONS
None.

RELATED AUTOCAD COMMAND
- 'Redraw   Redraws all viewports.

RELATED SYSTEM VARIABLES
None.

Dim: Restore

Rel. 1

Restores dimensions to the current dimension style.

Dim	Alt+	Menu bar	Alias	GC Alias
restore	...	...	d res	...

```
Dim: restore
Current dimension style: *UNNAMED
?/Enter dimension style name or RETURN to select dimension:
```

COMMAND OPTIONS
?   Lists dimension style names.
[Enter]   Selects a dimension with the style.

RELATED DIM COMMANDS
- Override   Overrides the current dimension style.
- Save   Saves a dimension style.

RELATED SYSTEM VARIABLE
None.

Dim: Rotated

Rel. 1

Draw linear dimension at any angle.

Dim	Alt+	Menu bar	Alias	GC Alias
radius	D,M,R	[Draw]	d ro	lx
		[Linear Dimensions]		
		[Rotated]		

```
Dim: rotated
Dimension line angle <0>:
First extension line origin or RETURN to select: [pick]
Second extension line origin: [pick]
Dimension line location (Text/Angle): t
Dimension text <>:
```

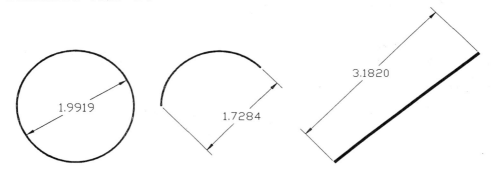

COMMAND OPTIONS
[Enter]   Displays submenu for dimensioning objects:
  Select line, arc or circle

RELATED DIM COMMANDS
- **Aligned**   Draws linear dimension aligned to two points.
- **Horizontal** Draws linear dimension horizontally.
- **Vertical**   Draws linear dimension vertically.

Dim: Save

Rel. 1

Saves dimension variables as a style.

Dim	Alt+	Menu bar	Alias	GC Alias
style	...	...	d sa	...

```
Dim: save
?/Name for new dimension style:
```

COMMAND OPTION
?   Lists names of dimstyles stored in drawing.

INPUT OPTION
~*dimvar*   (Tilde prefix) Lists differences between current and selected dimstyle.

RELATED AUTOCAD COMMAND
- DDim   Changes dimvar settings via dialogue box.

RELATED DIM COMMAND
- Status   Lists current status of dimension variables.

RELATED SYSTEM VARIABLES
- *All*
- DimStyle   Name of the current dimstyle.

Dim: Status

Rel. 1

Creates and edits dimstyles.

Dim	Alt+	Menu bar	Alias	GC Alias
status	...	...	d sta	...

Dim: **status**

Default values of all dimension variables:

DimVar	Default	Meaning
DimALT	Off	ALTernate units selected
DimALTD	2	ALTernate unit Decimal places
DimALTF	25.4000	ALTernate unit scale Factor
DimAPOST	...	Alternate text suffix (POST)
DimASO	On	create ASsOciative dimensions
DimASZ	0.1800	Arrow SiZe
DimBLK	...	arrow BLocK name
DimBLK1	...	1st arrow BLocK name
DimBLK2	...	2nd arrow BLocK name
DimCEN	0.0900	CENter mark size
DimCLRD	BYBLOCK	CoLoR of Dimension line
DimCLRE	BYBLOCK	CoLoR of Extension line and leader
DimCLRT	BYBLOCK	CoLoR of dimension Text color
DimDLE	0.0000	Dimension Line Extension
DimDLI	0.3800	Dimension Line Increment for continuation
DimEXE	0.1800	EXtEnsion above dimension line
DimEXO	0.0625	EXtension line Origin offset
DimGAP	0.0900	GAP from dimension line to text
DimLFAC	1.0000	Linear unit scale FACtor
DimLIM	Off	generate dimension LIMits
DimPOST	...	default suffix (POST) for dimension text
DimRND	0.0000	Rounding value
DimSAH	Off	Separate ArrowHead blocks
DimSCALE	1.0000	overall SCALE factor
DimSE1	Off	Suppress the 1st Extension line
DimSE2	Off	Suppress the 2nd Extension line
DimSHO	On	update dimensions while dragging (SHOw)
DimSOXD	Off	Suppress Outside eXtension Dimension
DimSTYLE	*UNNAMED	current dimension SYTLE (read-only)